Interplanetary Cubesats

Patrick H. Stakem

(c) 2017

3rd in the Cubesat Series

2nd edition

- Introduction ... 4
- The Author ... 5
- A note on Units ... 6
- Cubesats ... 6
- Challenges ... 7
- Technology Readiness Levels ... 8
  - Interplanetary Internet ... 10
- Special Cases – Interplanetary Cubesats ... 12
  - Communications ... 13
  - Constellations ... 14
  - Constellation support ... 15
- Where we're exploring ... 17
  - Exploring the Moon ... 18
  - Brazil's Garatea-L ... 21
- Exploring our solar system – the complexity of the Problem ... 21
  - Exploring the Sun ... 23
  - Exploring Mercury ... 25
  - Exploring Venus ... 25
  - Exploring Earth ... 27
  - Near Earth Objects ... 27
  - Exploring the Asteroid Belt, dwarf planets, and Centaurs ... 27
  - Exploring Comets ... 35
  - Exploring Mars ... 35
    - Marco Project ... 38
    - Saturn ... 39
  - Exploring the Ice Giants ... 41
    - Uranus ... 41
    - Neptune ... 42
  - Exploring Pluto, and beyond ... 42
- Making the Journey – How do we get there? ... 44
- Don't just send one ... 44
- Clusters ... 44
- Swarms ... 45
- Case Study – Swarms of Cubesats to Jupiter ... 46
  - The Cubesat alternative ... 46
  - Cubesats ... 49

- Ops concept 49
- Enabling Technologies 51
  - Better batteries and solar panels 51
  - Rad-Hard Software 51
  - Dispenser/Mothership 53
  - Inter-unit communications 53
  - Compute cluster of convenience – Beowulf 54
- Wrap-up 55
- References 56
  - Resources 59
- Glossary 61
- Country's that have launched Cubesats 76
- If you enjoyed this book, you might also enjoy one of my other books in the Space series. 78

To LEO, then Beyond...

Introduction

This book discusses the application of Cubesats in the exploration of our solar systems. Including the Sun, the eight primary planets and the minor planet Pluto, many moons, the asteroid belt, comets, the ring systems, the Centaurs, there is a lot to explore. Although the planets (and Pluto) have been visited by spacecraft, Earth's moon has been somewhat explored, and many of the other planets' moons have been imaged, there is a lot of "filling in the blanks" to be done. We examine the application of swarms of small independent spacecraft to take on this role. Some of the enabling technology's for cooperating swarms is examined.

Almost every Cubesat sent into space to this point has gone into Earth orbit, and is either there still, or has reentered the atmosphere. It's a big solar system, and there's a lot we don't know about it. Additionally, Cubesats are launched as ride-along payloads. There are two approaches for using Cubesats for exploration away from Earth. One uses the demonstrated technology of solar sailing, and missions using this approach are being implemented. Another uses a large carrier-mothership, loaded with hundreds of Cubesats. This is sent to a destination. achieves orbit, and dispenses the Cubesats, providing the communications link with Earth. JPL is postulating this type of mission in the 2020's. They baseline a dormant cruise duration of 100-2200 days, followed by a Cubesat life of 1-7 days. Prior to that, the most likely scenario is a traditional exploration mission with some tag-along Cubesats. The next step beyond that is to make a swarm of Cubesats the primary payload.

Missions to Mars and beyond and beyond are lengthy and expensive. We need to ensure that we are delivering payloads that will function and return new data. The trade off is between one or two large traditional spacecraft, or a new concept, a large number of nearly identical small spacecraft, operating cooperatively.

Necessarily, the Technology Readiness Level of this approach must be proven in Earth orbit, before the resources are allocated to deploy this approach to distant locations. Decades of time, and hundreds of millions of dollars are at stake. The application of Cubesats is seen to be the next revolution in Space Science.

NASA Plans ahead for Space Exploration based on the Decadal Surveys for Astrophysics, Heliophysics, and Planetary Science. This brings together the ideas of the science community at large to form a consensus on how to proceed. This is NASA's Science Mission Directorate reaching out to the Science community for direction. The latest report, available from the Planetary Society, is "Vision and Voyages for Planetary Science in the Decade 2013-2022" This report features Cubesat technology prominently. There is no guarantee that any of the projects will be funded.

Four important recommendations for the Cubesat community came from the most recent study. These are the Beyond LEO Smallsat Science Exploration Program, the Beyond LEO SmallSat Technology Maturation Program, the Small Spacecraft as Secondaries on All Beyond LEO Missions, and the Dedicated SmallSat Launch and Operations Program.

The Author

Mr. Stakem has been interested in rockets and spacecraft since high school. He received a Bachelor's degree in Electrical Engineering from Carnegie-Mellon University in 1971, where he was a member of the Applied Space Sciences group. His first job was for Fairchild Industries, then building the ATS-6 spacecraft. Wernher von Braun joined Fairchild as Vice-President of Engineering during this time, so, technically, Mr. Stakem was once a member of the von Braun Team. Specializing in support of spacecraft onboard computers, he has worked at every NASA Center. He supported the Apollo-Soyuz mission with the ATS-6 communications satellite. He received Master's degrees in Physics and Computer Science from the Johns Hopkins University. He has taught for Loyola University in Maryland, Graduate Department of Computer Science, AIAA, The Johns Hopkins University, Whiting School of Engineering, and Capitol Technology University.

He served for several years as a Mentor for Goddard Space Flight Center's Engineering bootcamp, which brought together international teams of students to work on real projects. This resulted in the deployment of the Grover (Greenland Rover), a large "satellite" operating at zero altitude. It collected data on the thickness of the Greenland Ice Sheet. He has also been active in a International Summer Cubesat program at Capitol Technology University, teaching classes and mentoring student projects.

This book is number 3 in the Cubesat Series, first published in 2017. The series includes Cubesat Engineering, Cubesat Operations, Cubesat Constellations, Clusters, and Swarms, and Interplanetary Cubesats. This current edition has been corrected and expanded with new material.

A note on Units

I am fairly conversant in both English and Metric units (what is the metric equivalent of furlongs per fortnight?). Metric (SI) is mandated for NASA usage now, for interchangeability with our partner space faring nations. Conversions are easy enough, but units conversion is a source of error. It's not what you know about units and measurement, its how you think.

Cubesats

A Cubesat is a small, affordable satellite that can be developed and launched by college, high schools, and even individuals. The specifications were developed by Academia in 1999. The basic structure is a 10 centimeter cube, (volume of 1 liter) weighing less than 1.33 kilograms. This allows several of these standardized packages to be launched as secondary payloads on other missions. A Cubesat dispenser has been developed, the Poly-PicoSat Orbital

Deployer, P-POD, that holds multiple Cubesats and dispenses them on orbit. They can also be launched from the Space Station, via a custom airlock. Most space-faring nations provide Cubesat launch services. The Cubesat origin lies with Professor Twiggs of Stanford University and was proposed as an approach to support hands-on university-level space education and opportunities for low-cost space access.

In what has been called the Revolution of smallsats, Cubesats lead the way. They represent paradigm shifts in developing space missions, opening the field from National-level efforts and large Aerospace contractors, to individuals and schools.

Challenges

Once we leave the vicinity of our home planet, conditions deteriorate quickly. The major issue is radiation, since we are outside of the trapped radiation belts, which provide some protection. This is a major challenge, but there are many known ways to mitigate this problem. Then, there is the thermal problem. We're going somewhere that's hotter (sun-ward), or colder. A big issue is the mission duration. It takes years to get to some of the outer planets, and even if the system is powered off, there can be events that will cause it not to wake up. Missions outside the rather friendly environment of near-Earth face additional challenges that must be addressed.

At the same time, communications becomes more difficult, and achievable data rates go down. The spacecraft might find itself on the other side of the Sun, from Earth, and communications would not be possible. JPL is exploring laser communication links for long distances.

Cubesats, being small, have constraints on power generation and storage, fuel storage (if any) and communications.

These are all solvable problems, but require additional engineering analysis.

Technology Readiness Levels

The Technology readiness level (TRL) is a measure of a device's maturity for use. There are different TRL definitions by different agencies (NASA, DoD, ESA, FAA, DOE, etc). TRL are based on a scale from 1 to 9 with 9 being the most mature technology. The use of TRLs enables consistent, uniform, discussions of technical maturity across different types of technology. We will discuss the NASA one here, which was the original definition from the 1980's.

Technology readiness levels in the National Aeronautics and Space Administration (NASA)

1. Basic principles observed and reported

This is the lowest "level" of technology maturation. At this level, scientific research begins to be translated into applied research and development.

2. Technology concept and/or application formulated

Once basic physical principles are observed, then at the next level of maturation, practical applications of those characteristics can be 'invented' or identified. At this level, the application is still speculative: there is not experimental proof or detailed analysis to support the conjecture.

3. Analytical and experimental critical function and/or characteristic proof of concept.

At this step in the maturation process, active research and development (R&D) is initiated. This must include both analytical studies to set the technology into an appropriate context and laboratory-based studies to physically validate that the analytical predictions are correct. These studies and experiments should constitute "proof-of-concept" validation of the applications and concepts formulated at TRL 2.

4. Component and/or breadboard validation in laboratory environment.

Following successful "proof-of-concept" work, basic technological elements must be integrated to establish that the "pieces" will work together to achieve concept-enabling levels of performance for a component and/or breadboard. This validation must be devised to support the concept that was formulated earlier, and should also be consistent with the requirements of potential system applications. The validation is "low-fidelity" compared to the eventual system: it could be composed of ad hoc discrete components in a laboratory

TRL's can be applied to hardware or software, components, boxes, subsystems, or systems. Ultimately, we want the TRL level for the entire systems to be consistent with our flight requirements. Some components may have higher levels than needed.

5. Component and/or breadboard validation in relevant environment.

At this level, the fidelity of the component and/or breadboard being tested has to increase significantly. The basic technological elements must be integrated with reasonably realistic supporting elements so that the total applications (component-level, sub-system level, or system-level) can be tested in a 'simulated' or somewhat realistic environment.

6. System/subsystem model or prototype demonstration in a relevant environment (ground or space).

A major step in the level of fidelity of the technology demonstration follows the completion of TRL 5. At TRL 6, a representative model or prototype system or system - which would go well beyond ad hoc, 'patch-cord' or discrete component level breadboarding - would be tested in a relevant environment. At this level, if the only 'relevant environment' is the environment of space, then the model/prototype must be demonstrated in space.

7. System prototype demonstration in a space environment.

TRL 7 is a significant step beyond TRL 6, requiring an actual system prototype demonstration in a space environment. The prototype should be near or at the scale of the planned operational system and the demonstration must take place in space.

The TRL assessment allows us to consider the readiness and risk of our technology elements, and of the system.

Although Cubesats themselves are at TRL-7 for Earth orbit, the concept of Cubesat Swarms and Clusters are at TRL levels 3-4. Cubesat missions are currently heading to the Moon and Mars.

Interplanetary Internet

Communications between planets in our solar system involves long distances, and significant delay. New protocols were needed to address the long delay times, and error sources.

A concept called the Interplanetary Internet uses a store-and-forward node in orbit around a planet (initially, Mars) that burst-transmits data back to Earth during available communications windows. At certain times, when the geometry is right, the Mars bound traffic might encounter significant interference. Mars surface craft communicate to Orbiters, which relay the transmissions to Earth. This allows for a lower wattage transmitter on the surface vehicle. Mars does not (yet) have the full infrastructure that is currently in place around the Earth – a network of navigation, weather, and communications satellites.

For satellites in near Earth orbit, protocols based on the cellular terrestrial network can be used, because the delays are small. In fact, the International Space Station is a node on the Internet. By the time you get to the moon, it takes about a second and a quarter for electromagnetic energy to traverse the distance. Delay tolerant protocols developed for mobile terrestrial communication were used, but break down in very long delay situations.

We have a good communications model and a lot of experience in Internet communications. One of the first implementations for space used a File Transfer Protocol (FPP) running over the CCSDS space communications protocol in 1996.

The formalized Interplanetary Internet evolved from a study at JPL, lead by Internet pioneer Vint Cerf, and Adrian Hook, from the CCSDS group. The concepts evolved to address very long delay and variable delay in communications links. For example, the Earth to Mars delay varies depending on where each planet is located in its orbit around the Sun.

The Interplanetary Internet implements a Bundle Protocol to address large and variable delays. Normal IP traffic assumes a seamless, end-to-end, available data path, without worrying about the physical mechanism. The Bundle protocol addresses the cases of high probability of errors, and disconnections. This protocol was tested in communication with an Earth orbiting satellite in 2008.

NASA's Goddard Space Fight Center in Greenbelt, Maryland, has been the hub of the space data network since the beginning. In the Apollo era, a world-wide system of ground stations providing continuous coverage was not yet in existence. NASA supplemented their ground stations with a series of tracking ships, to fill in coverage gaps. All data came to the basement of the Operations Building, 14, at Greenbelt. It was then routed upstairs to satellite control centers, or to Houston or Marshall for Manned flights. For its interplanetary missions, JPL maintained the Deep Space Network, a set of three very large antennas spaced around the world. During launch and near Earth operations, these were supplemented by NASA's world-wide set of tracking stations for Earth orbiting satellites.

Special Cases – Interplanetary Cubesats

The Marco Mission to Mars was launched in 2017, with a pair of tag-along Cubesats and the primary payload. Several mission to the Moon for Cubesat Platforms are in work.

NASA's Planetary Cubesat Science Institute (PCSI) is focusing on developing Cubesat approaches for high priority science. In August of 2016, they held a symposium at Goddard Space Flight Center, getting key people together to exchange ideas. Some pending projects include the Hydrogen Albedo Lunar Orbiter (HALO) and the Primitive Object Volatile Explorer (Prove), which will do a close fly-by of the comet 46P at its perihelion in December 2018, sampling the volatiles that are boiling off.

NASA/GSFC set up a Cubesat Initiative as part of its Science Mission Directorate in 2013. They saw the advantages to low cost and rapid development that Cubesats can offer.

NASA, particularly Jet Propulsion Lab, is defining new approaches to exploration away from Earth's neighborhood, utilizing Cubesat technology. Many missions using solar sails are being considered, and the technology has been demonstrated. Up to this point, the application of Cubesats for interplanetary exploration has been approached by building bigger, more robust Cubesats. Later, we will discuss another approach involving standard launch vehicles, carrying large numbers of Cubesats.

As we get farther from Earth, the Cubesat's small antennas, and relatively low power, means we have to get clever with communications. There will be a limited bandwidth. This happened with the New Horizon's spacecraft at Pluto – It took more than 16 months to transmit all the data back. JPL's new approach will use laser communications.

At the moment, there is a communications relay satellite in Mars orbit. This is for the Mars rovers on the surface. We can postulate a mothership can be placed into orbit around the target, and provide this service for the Cubesats. JPL is also exploring onboard processing of the data, to a point. This involves, then, a computation-communication trade. Outward, away from Earth, power from the Sun is problematic. When Mars and Earth are in opposition, on opposite sides of the Sun, communication is not possible. There is a Communications blackout. This is true for all of the planets. This can span a series of days or weeks. Adverse space weather can also affect interplanetary communications.

NASA's involvement in this area comes from the Advanced Exploration systems Division. Planned missions include BioSentinel, with living organisms (yeast) onboard. The mission will study the radiation effects on their DNA, for a period of 18 months.

An important mission is the Near Earth Asteroid Scout. It will use solar sail propulsion, and target an asteroid about 90 meters in diameter to study.

Communications

NASA's Deep Space Network consists of three sites spaced around our planet. It supports deep space missions for NASA and other entities. It is managed by the Jet Propulsion Lab (JPL) in Pasadena, California. The nearest station to JPL is at Goldstone, in the desert to the east. Two other stations, in Spain and Australia are spaced about 120 degrees apart on the globe from Goldstone. The DSN started operations in the 1960's, with teletype communications with the Pasadena facility. The DSN is heavily oversubscribed, supporting numerous deep space missions.

Several approaches to communication with spacecraft at a large distance from Earth, and examining other planets have been defined. The Interplanetary Internet implements a Bundle Protocol to address large and variable delays. Normal IP traffic assumes a seamless, end-to-end, available data path, without worrying about the physical mechanism. The Bundle protocol addresses the cases of high probability of errors, and disconnections. This protocol was tested in communication with an Earth orbiting satellite in 2008.

Constellations

Constellations are groups of satellites operating together to observe a single target A constellation allows you to do simultaneous observations from multiple locations. The elements of a constellation can be homogeneous or not. With Cubesats, we would have multiple co-operating units. There can be distributed control, or central control.

NASA says, " A Constellation is a space mission that, beginning with its inception, is composed of two or more spacecraft that are placed into specific orbit(s) for the purpose of serving a common objective (e.g., Iridium)"

An example is their Distributed Spacecraft Mission being defined at JPL to allow "formation flying" of multiple spacecraft. Another point of view is the "fractionalized spacecraft", where the spacecraft functionality is distributed across multiple units. Critical to this is a intra-communications mechanism. The Constellation may use a mesh or lattice architecture. The members of this organization can be launched together, or separately. One thing for sure, each unit needs a unique identifier.

These units can be statically or dynamically allocated. We might have a fixed plan for location and function, or it might be ad hoc, responding to conditions as they are encountered.

For Cubesats, we can use a "mothership" that takes the members to their destination, then sticks around to manage. This approach will be discussed in a section on the Strawman multi-Cube sat mission to Jupiter, later in the book.

Constellations add complexity to the mission. Complexity has two major components, the number of units, and the number of interactions among members.

Constellation support

In some cases, the mission may involve a constellation of multiple spacecraft. This includes the GPS constellation, the weather satellite systems TDRSS, and a number of commercial communications satellites, providing data service world-wide, as well as Cubesats.

Managing a constellation adds to the complexity. Even if each spacecraft is built to the same plan, different spacecraft, launched at different times, and having differing times on-orbit, need customized attention. The most important aspect is to have a unique identifier, so you know which spacecraft you're talking to.

An approach to Constellation control centers can involve a hierarchy of a master control center and with multiple space assets to control, or a peer network of individual control centers, that also provides a built-in redundancy and backup. A backup control center is useful not only for anomalies at the primary center, but also to allow for maintenance and upgrade of the primary center, and for personnel training and certification.

A major constellation is NOAA's Polar and Geostationary weather and environmental satellites. The USAF operates the Global Positioning System (GPS) Constellation.

Constellations of communication satellites are used for commercial ventures such as DishNetwork, a satellite TV provider, and Iridium and GlobalStar, communications constellations.

The USAF maintains a world-wide satellite control network. The 2nd Space Operations Squadron in Colorado is typical of the units involved in constellation operations. The system has been in existence for decades, starting in the mainframe era of the 1970s, transitioning to the client-server architecture, and have been modernized to pc and server architecture. Extensive training for operators is provided.

An ongoing debate in the optimum architecture for multi-satellite control is between a centralized design, and a distributed architecture. Centralized is the legacy approach. Distributed takes advantage of advances in networking and abstraction. In the distributed approach, multiple ground stations and control centers are linked by existing terrestrial data communication resources.

The distributed architecture scales more freely, with computation, storage, and communications resources being added as demand increases. High system reliability and security can be maintained from industry best practices. The scale-able, distributed technology has been driven by large data-centric organizations such as Google, and retailers such as Amazon, as well as social media sites such as Facebook and U-tube. These do not meet military-grade security, of course, but that can be addressed.

Another advantage of the distributed approach is dynamic allocation of resources, having (and paying for) resources when you need them, not all the time. The system provides mission safety simultaneously with cost effectiveness. A metric of interest is the staff to spacecraft ratio. If domain-skilled staff can be shared among the constellation, yet be brought together in the case of anomalies, personnel costs can be contained. Distributed approaches give economy of scale.

The same information for each spacecraft in a homogeneous constellations provides summaries of critical cross-platform information. If we just had a failure on one spacecraft, we will look for that to happen on others. A merged database, allows for trending information to flow forward. As constellations age, the individual members age and fail at different rates. From trending data on early failures, the remaining spacecraft can be monitored especially for known failures and degradation.

Where we're exploring

This section discusses the requirements for missions to the other planets of our solar system. Each represents unique challenges. Earth's moon is the only extraterrestrial body to be visited by Man (so far). The amount of diversity among the planets, their associated moons, ring systems, comets, asteroids, and other strange items is staggering.

We have had a closer look at all other bodies in our solar system, mostly by fly-by's, and have landed rovers on several. Probes have gone to examine 13 minor planets, asteroids, and dwarf planets as well.

NASA's SIMPLEx mission (Small Innovative Missions for Planetary exploration) has 13 Cubesats lined up for a 2018 launch on an Orion vehicle. This will spend three weeks in space, including 6 days orbiting the moon. Cubesat missions will include the Lunar Flashlight, to look for water ice; the Near-Earth Asteroid Scout (with a solar sail); the BioSentinal, SkyFire, from Lockheed Martin; Lunar Ice Cube; CUSP, Cubesat for Solar Particles, and Lunar Polar Hydrogen Mapper. More will be selected closer to launch time.

The Simplex program received responses to its Request for Proposals for Interplanetary Missions in 2015. They chose to fund two missions, the Lunar Polar Hydrogen Mapper, a 6-U cubesat to be placed in lunar polar orbit, and Q-Pace (Cubesat Particle Aggregation and Collision Experiment), which will study low velocity particle collisions in microgravity.

The program also selected some additional missions for technology development. These include the Mars Micro orbiter (MM) which flys a 6U Cubesat to observe the Martian atmosphere in the visible and infrared.

The hydrogen albedo Lunar orbiter (Halo) is a 6U Cubesat with a propulsion system looking to find the origin of water on the lunar surface. This is a GSFC mission. There is known water ice at the lunar poles.

Last, and least, is the Diminutive Asteroid Visitor using Ion Drive (DAVID) which is a 6U mission to visit an Earth orbit crossing asteroid. This is a project of NASA-Glenn.

We will not cover Earth exploration by Cubesat, as that is an ongoing effort. We focus on visiting our neighbors.

Exploring the Moon

The Moon was the first extra-terrestrial body to be explored by humans. It is close enough that the communication time is about ½ second, and lunar spacecraft could be controlled from Earth. But, the lack of communication with the craft when they are behind the moon, from the Earth viewpoint, dictated at least a stored telemetry and command capability. Rovers on the face of the moon towards Earth are in continuous contact.

Early in the era of space exploration, a series of rover vehicles were sent to the Earth's moon. These were designed as precursors to a manned visit. From the mid-1960's through 1976, there were some 65 unmanned landings on the moon. This was also the subject of a private effort, the Google X-prize. The moon is still the subject of intense study, with missions from the United States, Russia, China, India, the European Union, and Japan.

The Soviet Union launched a series of successful lunar landers, sample return missions, and lunar rovers. The Lunokhod missions, from 1969 through 1977, put a series of remotely controlled vehicles on the lunar surface.

The NASA Surveyor missions of 1966-68 landed seven spacecraft on the surface of the moon, as preparation for the Apollo manned missions. Five of these were soft landings, as intended. All of these were fixed instrument platforms.

Yutu is the name of the Chinese Lunar Rover, and means Jade Rabbit. It was launched in December of 2013. It landed successfully on the moon, but became stationary after the second lunar night. It is a 300 pound vehicle with a selection of science instruments, including an infrared spectrometer, 4 mast-mounted cameras including a video camera, and an alpha particle x-ray spectrometer. The rover is equipped with an arm. It also carries a ground penetrating radar. It is designed to enter hibernation mode during the 2-week lunar night. It does post status updates to the Internet, and still serves as a stationary sensor platform.

The latest mission to study the Moon is the Lunar Reconnaissance Orbiter, LRO, from NASA/GSFC. It was launched in 2009, and is still operating. It is is a polar orbit, coming as close as 19 miles (30 km) to the lunar surface. It is collecting the data to construct a highly detailed 3-D map of the surface. Up to 450 Gigabits of data per day are returned to Goddard.

One interesting feature on the moon is the lava tubes, long channels left after lava flow. They exist on the Earth as well, and Mars. These might be exploited as lunar habitats, or shelters in case of radiation storms.

Since the moon is in tidal lock, we never see the backside from Earth. So close, yet so far away. Any rover on the lunar backside is permanently out of touch with Earth, because the moon is in tidal lock, and one side always faces Earth. Lunar orbiters are out of communications with Earth for somewhat less than a half orbit. This could be solved with a communications relay satellite in lunar polar orbit. The moon does wiggle a bit, less than 50% of the surface is permanently out of touch.

If we could adjust the lunar comm relay spacecraft's orbit so it remains normal to the Earth-Moon axis, it would be ideal. This might involve excessive propellant expenditure, shortening the mission life.

A lunar orbiter can store data on the back side, and send it back to Earth when it's over the front side. Might be a nice job for a Cubesat. As a student project during the Summer of 2016, the author mentored an engineering team to develop a Cubesat-based rover for explanation of the lunar backside. This was a ground-based tracked vehicle. It had an associated lander/habitat where it could spend the 14-day lunar night. This prototype was tested at NIST's robotic testing lab, which usually sees urban search and rescue and bomb disposal robots.

An interesting NASA Cubesat mission, the Lunar Flashlight, involves a 6U Cubesat with a solar sail. It will go into lunar orbit, and look for ice deposits and potential areas of resource extraction (ie, lunar mining areas). It will be able to use it's solar sail to reflect sunlight onto the surface, particularly the polar regions. The solar sail will be 80 square meters in size.

The first Artemis mission to the Moon will carry along 13 6-U Cubesats. These include ArgoMoon from the Italian Space Agency, BioSentinel, Cubesat for solar particles from Southwest Research Institute, Equuleus, from the Japanese space agency, Lunar Flashlight, from Morehead State University (Kentucky), Lunar Polar Hydrogen Mapper from Arizona State, Near-Earth Asteroid Scout, with a solar sail, Omotenashi, also from the Japanese space agency, and Lunir, from Lockheed Martin., Cislunar Explorers, from Cornell University, Earth Escape Explorer, univeristy of Colorado Boulder, and Team Miles, from Fluid and Reason, Tampa Florida.

The first Artemis mission, uncrewed, is scheduled for December 2021.

Brazil's Garatea-L

All over the world, any nation can become spacefaring with a reasonable cost. This ambitious project, to be launched in 2020, is an example. It will orbit the moon, and examine the effects of that environment on different life forms. The project is being done in cooperation with the UK Space Agency and ESA. Several other Cubesats will tag along for the ride. The Brazilian mission will provide the communications back to Earth for all of the associated missions. The launch will carried out by the European Space Agency (ESA). The UK Space Agency on the same flight will send the Pathfinder, the first commercial deep space mission.

It is a joint effort among top Brazilian universities: The National Institute of Space Studies (INPE), the Technological Institute of Aeronautics of USP, the National Laboratory of Synchrotron Light and the Institute of Technology of the Pontifical Catholic University of Rio Grande do Sul, with the University of Sao Paulo in the lead.

Garatea, in the native Tupi-Guarani language means "looking for lives."

A Cubesat mission to the Earth-Moon Lagrange point 2 (behind the Moon) would give us a good idea of whether it would be a good place for a radio astronomy observatory. There, the Moon shields the spacecraft from radio noise from Earth. The concept would involve a number of Cubesats, flying in formation, and forming a large virtual radio telescope.

This section will discuss some of the other objects in our solar system, and how they have been explored, with discussions of the environments.

Exploring our solar system – the complexity of the Problem.

Let's look at the number of objects in our solar system that we would like to know more about. We will also list the one-way light times for the various objects. This tell us how long a radio signal takes to traverse that distance.

Trojan Asteroids are in the same orbit as a primary, in front and behind (at the 4th and 5th Lagrange points). Jupiter has the most, at around 600,000. The cluster at L4 is called the Greek Camp, and at L5 is the Trojan Camp. Trojan 617 Patroclus in the Jovian system is a binary object.

Earth has 1 moon and one Trojan. Earth's moon is about ½ light second away. There are more than 15,000 Near-Earth Objects (NEO's). Technically, an NEO is a solar system object whose closest approach to the Sun is 1.3 AU, and that comes in close proximity to the Earth There are 14,000 known asteroids in this category, 100 comets, solar orbiting spacecraft, and meteoroids. All these have the potential of striking the Earth. They are closely tracked from the ground, by NASA's Planetary Defense Coordination Office. A joint US/EU project called Spaceguard is tracking NEO's larger than 30 meters. Three NEO's have been visited by spacecraft. Cubesats, with solar sails, are an ideal approach to explore objects in our home vicinity, because of the danger they may present, but also for possible exploitation by mining.
Mercury has no moons. The planet is in tidal lock with the Sun. It wiggles a bit, creating a twilight zone where the temperature extremes are not as bad.

Venus has no moons, but does have Trojans. The planet is trapped in a thick cloud mantel, and is extremely hot at surface, due to the Greenhouse effect.

The asteroid belt includes Ceres the dwarf planet, and 750,000 rocks larger than 1 km. It is a torus, extending from about 2.0 AU to 3.2 AU..

Mars has 2 moons and 7 Trojans. The one-way light time varies between 3 and 29 minutes. It has an infrastructure in orbit for relaying surface rover's data back to Earth, and to observe weather to predict surface sand storms.

The Gas Giants are two large planets, Jupiter and Saturn, beyond the orbit of Mars. They consist mostly of the gases hydrogen and helium. At depth, they consist of liquid metallic hydrogen. There may be a core of solid hydrogen, or a rocky material. It is possible these elements exist as a liquid at extreme depths. Essentially, a gas giant planet is a failed star. The nuclear fusion process did not get started.

Jupiter has 79 known moons, and perhaps 1 million Trojans of 1 kilometer or larger. These tend to congregate at the L4 and L5 points. The largest has a diameter of several hundred kilometers. The International Astronomical Union just announced as this book was being updated the discovery of 12 previously unknown moons of Jupiter, by an observatory high in the Andes in Chile. Only one has been named so far, Valetudo, a great-granddaughter of Jupiter

The one way light time for Jupiter to Earth is 33-53 minutes.

Saturn and it's 62 known moons has a one-way light time to Earth around 1.4 hours. Saturn has been visited by spacecraft four times. The first was a flyby by Pioneer-11 in 1979. This showed the temperature of the planet was 250 degrees K. Saturn missions tend to go by way of Jupiter, and use it's gravity for an extra boost in velocity.

Exploring the Sun

We get all our energy from this huge thermo-nuclear reactor some 8 light-minutes away. As you get closer to the Sun, its gets hotter, and there is more radiation in terms of energetic particles.

NASA sent a series of probes to observe the sun in 1958 and for ten years after. These were the Pioneer 5-9 spacecraft. They didn't get any closer, but provided a different point of view. They got good data on the solar wind and the sun's magnetic field. Pioneer 9 sent back good data for 15 years. The Helios spacecraft in the 1970's were joint U.S.-German missions that used an orbit that got within the orbit of Mercury.

The 1980 Solar Maximum Mission observed the Sun in the spectrum of gamma rays, X-rays, and Ultraviolet. SMM had a failure in its electronics months after launch, but was repaired by a subsequent Shuttle mission. SMM used the NASA Standard Spacecraft Computer (NSSC-1) constructed of discrete logic elements. The author had flight software onboard SMM, when it reentered the atmosphere and burned in 1989.

A Japanese Mission to study the Sun was Yohkoh, or Sunbeam, in 1991. It imaged solar flares in the x-ray spectrum.

Other US missions included SOHO, the Solar and Heliospheric Observatory, and the Solar dynamics Observatory. SOHO was located at the Lagrangian point between the Sun and the Earth, while is a null point in the gravity field. It sees the Sun constantly in many selected wavelengths.

The Solar Terrestrial Relations Observatory (Stereo) is a dual spacecraft mission to the Sun, launched in 2006. One is ahead of the Earth in orbit, the other behind. This gives three points of view of solar phenomena.

The Ulysses spacecraft, discussed in the section on Jupiter, left the plane of the ecliptic (thanks to the Jupiter swing-by) and observed the Sun's high latitudes. One of its discoveries was that large magnetic waves emitted from the Sun scattered galactic cosmic rays.

The Genesis mission was designed to capture and return solar material. It achieved its goal, but was damaged in a crash landing when it returned to Earth in 2004.

Two areas that have never been imaged are the Sun's polar regions. This is because of the very large energy expenditure required to get out of the plane of the ecliptic. This would be a good mission for a Cubesat, or a few Cubesats. The process of achieving an out-of-ecliptic plane and into a path that gets us to the solar poles continues as a research topic.

A proposed mission, the Solar Polar Constellation, would be dedicated to high inclination solar orbit.

The latest solar mission, just launch, the Parker Solar Probe, is scheduled to head out, around the time this book is published. Ir is to fly into the lower solar corona. It has a large composite heat shield. It will make seven Venus fly-bys on its way. A highly elliptic orbit around the Sun will be used, to allow the spacecraft to cool off between observations.

In addition, Cubesats missions could be used to study the Solar Wind at various distances from the sun, using that same wind and solar sails they use to get around. It would be nice to have additional monitoring platforms between Earth and the Sun to give us earlier warning of Solar storms

Exploring Mercury

The U. S. Messenger mission to Mercury, the closest planet to the Sun, was launched in 2004. It is currently orbiting the hottest planet. No landing on Mercury has been attempted, although it would be feasible in the *twilight zone* between the extremely hot solar facing side, and the much colder space facing side. Mercury is in tidal lock with the sun, with one side always facing it. It wobbles a bit, creating a "twilight zone" that is much less extreme. It has no known moons, or Trojans. Being so close to the Sun, it is difficult to observe the planet and its immediate vicinity.

Exploring Venus

The Soviet space program sent a series of probes to Venus. Early efforts were either crushed in the dense atmosphere, or suffered thermal damage. The Venera-7 mission had a goal of surface sample return. It struck the surface harder than planned, but returned temperature data for about 20 minutes. The Venera-8 probe returned data for some 50 minutes. Venera-13 and -14 returned color photos of the surface. Further Soviet and US efforts involved observation from Venus orbit. The Venus environment has proven extremely hostile. It seems our sister world, next towards the Sun from us, is in a environmental runaway condition. Heavy greenhouse clouds trap the solar energy, and cause massive global warming on a planetary scale. The surface temperature is high enough to melt some metals. This is very hard on computers, and electronics in general.

Venus' atmosphere is 96% carbon dioxide at a surface pressure of nearly 100 times Earth's, a greenhouse gone wild. It has no moons. Venus is roughly Earth-sized, but something went terribly wrong. It also has clouds of sulphuric acid, that landers have to get through. There is no magnetic field, but there is active volcanism.

Venus Express, an ESA mission, is in Venus Polar orbit. It found a massive double atmospheric vortex (storm) at the south pole. Venus Express operated from 2005-2014. Venus has no moons, but does have Trojans. The Japanese Venus Climate Orbiter "Akatsuki" was launched in 2010, but failed to achieve Venus orbit. It orbited the Sun for 5 years, and was finally put into Venus orbit in 2015.

JPL is working on a Cubesat mission to Venus, it will have a mission duration of 30 days. It is to sample the noble gases using an ion-trap mass spectrometer in the atmosphere.

CUVE is a 12-U cubesat mission to explore Venus in the Ultraviolet. This is being done for NASA/GSFC by the University of Maryland. It will have a multi-spectral UV imager, a high resolution ultraviolet spectrometer, and small UV telescope.

Rocket Labs, of California, is designing a new satellite bus, named Photon. This will be launched to Venus in 2023. It will support a laser-tunable mass spectrometer in the Venus atmosphere,

Exploring Earth

There are hundreds of Cubesats in orbit around Earth, and some on the International Space Station. In 2018 there will be a Cubesat mission to the moon. Of interest to Cubesat exploration will be the Near Earth Objects (NEO), or Near Earth Asteroids (NEA). Another application is studying the moon's backside, the side that always faces away from Earth. Lunar orbits can be low, since there is essentially no atmosphere. You only have to clear the mountains. The back side of the moon is also an ideal location for radio telescope, being shaded from the Earth's RF clutter.

Near Earth Objects

Technically, an NEO is a solar system object whose closest approach to the Sun is 1.3 AU, and that comes in close proximity to the Earth There are 14,000 known asteroids in this category, 100 comets, solar orbiting spacecraft, and meteoroids. All these have the potential of striking the Earth. They are closely tracked from the ground, by NASA's Planetary Defense Coordination Office. A joint US/EU project called Spaceguard is tracking NEO's larger than 30 meters. Three NEO's have been visited by spacecraft.

Cubesats, with solar sails, are an ideal approach to explore objects in our home vicinity, because of the danger they may express, but also for possible exploitation by mining.

Exploring the Asteroid Belt, dwarf planets, and Centaurs

Asteroids have been imaged by the New Horizons spacecraft, on its way to Pluto, and by the Cassini spacecraft. The Pioneer-10 spacecraft was sent to study the far reaches of the solar system It passed through the Asteroid belt on its way to Jupiter and Saturn, and collected valuable data.

A driver in the space environment is the exploration of the asteroids, numbering in the thousands. Although there are fewer than 10 planets, and less than 200 moons, there are millions of asteroids, mostly in the inner solar system. The main asteroid belt is between Mars and Jupiter. Each may be unique, and some may provide needed raw materials for Earth's use. There are three main classifications: carbon-rich, stony, and metallic.

The physical composition of asteroids is varied and poorly understood. Ceres appears to be composed of a rocky core covered by an icy mantle, whereas Vesta may have a nickel-iron core. Hygiea appears to have a uniformly primitive composition of carbonaceous chondrite. Many of the smaller asteroids are piles of rubble held together loosely by gravity. Some have moons themselves, or are co-orbiting binary asteroids. The bottom line is, asteroids are diverse.

It has been suggested that asteroids might be used as a source of materials that may be rare or exhausted on earth (asteroid mining) or materials for constructing space habitats or as refuelling stations for missions. Materials that are heavy and expensive to launch from earth may someday be mined from asteroids and used for space manufacturing. Valuable materials such as platinum may be returned to Earth for a profit.

A Cubesat mission, AIDA (Asteroid Impact and Deflection) is a pair of satellitesto study asteroid deflection. It was opriginally part of an ESA mission. It has been downsized, but will impact the asteroid 65803 Didymos in 2022. Five years after that, another spacecraft named Hera will photograph the impact point.

There are only 8 ½ planets, but there are thousands of asteroids, and it seems there may be as many types. This means that exploring the known asteroids is a daunting challenge. On the other hand, the asteroids can be a significant source of raw materials for Earth. A conventional survey and exploration approach would take too long. What is needed instead is a multitude of autonomous and flexible nano-spacecraft. The architectural model is a swarm (social insect model) distributed intelligence. The platform of low cost, low power, low weight could involve Cubesats with solar sails. The asteroid belt contains Ceres, the Dwarf planet, and some 750,000 rocks larger than one kilometre in diameter.

The physical composition of asteroids is varied and poorly understood. Ceres appears to be composed of a rocky core covered by an icy mantle, whereas Vesta may have a nickel-iron core. Hygiea appears to have a uniformly primitive composition of carbonaceous chondrite. Many of the smaller asteroids are piles of rubble held together loosely by gravity. Some have moons themselves, or are a co-orbiting binary pair. The bottom line is, asteroids are numerous and diverse.

The asteroids are not uniformly distributed. In the asteroid belt, the Kirkwood gaps are relatively empty spots. This is caused by orbital resonance of the asteroids with Jupiter. Orbiting irregular shaped bodies is challenging, due to the irregular gravity field. This makes station keeping and attitude control a problem.

A Centaur is a type of dwarf planet, not quite making the cut to "real planet." there are 44,000 known examples with diameters greater than 1 km. They have unstable orbits that intersect those of the gas giants. They are somewhat like asteroids, and somewhat like comets. The largest known, Chariklo, has a ring system. These have not been photographed from a close position.

A intelligent swarm solution to resource exploration of the asteroid belt was proposed by Curtis, et al, in 2003. The concept of Cubesats were not enough advanced for the authors to specifically mention them. The project implements the NASA ANTS (Autonomous Nano-Technology Swarm concept (Truszkowski, et al) showing division of labor, colloboration, and reliability through redundancy. They mention that members of the swarm could occasionally engage in risky behavior, since they were many redundant units (take one for the team). They baselined about 1,000 units. They defined 8-10 types of *Workers,* each with specific capabilities. Units assigned to swarm cohesion and communication they term *Messengers*. There are also Rulers, who function in a managerial role. Cubesats are not specifically mentioned, but the application is certainly feasible. This is a swarm.organized in a hierarchy. A distributed control approach is also feasible.

It has been suggested that asteroids might be used as a source of materials that are rare or exhausted on earth (asteroid mining) or materials for constructing space habitats or as refueling stations for missions. Materials that are heavy and expensive to launch from Earth will someday be mined from asteroids and used for space manufacturing. Valuable materials such as platinum may be returned to Earth for a profit.

Exploring the asteroids requires a diverse and agile system. Thus, a swarm of small spacecraft with the same architecture but different capabilities can be used be used, combining into Teams of Convenience to address situations and issues discovered in situ. The Cubesats would be carried to the vicinity of the target environment in a "Mothership," larger, but with the same basic architecture. It is, in a sinse, a big Cubesat. It would provide propulsion, station-keeping, communications relay, electrical charging, and deployment of explorer Cubesats. The Cubesats are the primary payload, not tag-alongs.

For the Asteroid belt, a payload of 300, 6U Cubesats would not be unreasonable.

Each member of the swarm will be aware visually of other swarm members in close proximity. This will be facilitated by having the Mothership as the center of the coordinate system. It will determine its position by celestial navigation. The Cubesats will have a similar capability. The mothership will maintain, as part of its onboard database, the location of all other members. It will also monitor for pending collisions and warn the participants. There will be rules concerning how close swarm members can get to each other, a virtual zone of exclusion. All Earth-based interaction with the swarm will be through the Mothership. Due to varying communication delays, operation of the swarm by tele-operation from Earth is not feasible. The Swarm could be on the opposite side of the Sun from the Earth for extended periods. This is addressed by building autonomy into the system, and a large amount of non-volatile storage will be included for science data.

Biological swarms, such as ants, achieve success by division of labor throughout the swarm, collaboration, and sheer numbers. They have redundancy, as any individual can do any task assigned to the swarm. The individual units are highly autonomous, but are dependent on other members for their needs. They achieve success with a simple neural architecture and primitive communications.

Each unit will have on its external surface a plaque with a unique QR code, enabling visual identification of units. Each swarm member will be equipped with one or more cameras, not only for target investigation, but also for observing the position and relative motions of other swarm members.

The Cubesats would follow NASA/GSFC's Pi-Sat architecture. Using standard linux clustering software (Beowulf), the Mothership and undeployed swarm members will be able to form an ad-hoc cluster computer to process science data in-situ. Within the Mothership, a LAN-based Mesh network software will be used. The Mothership's main computer will be a Raspberry-Pi based cluster.

At the distance of the asteroid belt, the solar constant (kw/meter^2) is about ¼ of that at Earth's distance from the sun. The Cubesats would be launched from the mothership fully charged, and could have a small solar array. A 30-day life time would be satisfactory. The cubesats are not recovered.

The mothership porvides cloud services to the deployed units of the swarm. It would also have a bi-propellant engine for orbit and cruise adjustments, and a monopropellant system for attitude control and reaction wheel momentum unloading.

In Swarm robotics, the key issues are communication between units, and cooperative behavior. The capability of individual units nodes not much matter; it is the strength in numbers. Ants and other social insects such as termites, wasps, and bees, are models for robot swarm behavior. Self-organizing behavior emerges from decentralized systems that interact with members of the group, and the environment. Swarm intelligence is an emerging field, and swarm robotics is in its infancy. Co-operative behavior, enabled by software and intra-unit communications has been demonstrated.

They define target selection according to mission goals, but also note that mission goals change as data is collected at the site. The concept of multiple spacecraft coming together to form virtual instruments is discussed. Here, we might have implement observations from multiple points.

The Operational Concept involves teams that produce data and some higher level products, which are communicated to Messengers, and archived. The Rulers oversee data flow. When a sufficient amount is collected, a Messenger will be dispatched to carry it back (Today, this could be accomplished with radio or laser link).

A surface lander/hoverer could also be included in the Cubesat suite. Standard space communications protocols will be used between the lander and the Mothership, via UHF link.

The Cubesat members will collect observation data on their target, and can conduct radio occultation experiments to better categorize the distribution of particles. They can also conduct synchronized simultaneous observation from multiple observation points of features of interest.

The Mothership is the navigation reference point for the Cubesats. It obtains its position with respect to Earth from observation, and ground tracking. There will be times when the Earth is not visible form the Mothership's position, so it will use extrapolation and local observation. During these periods of occultation, and also periods of long one-way light times, the Mothership assumes local responsibility for the Health and Safety of the Swarm members. For this, we will implement Control Center functionality within the Mothership. This will take the form of Ball Brother's COSMOS software. This product addresses traditional system test, integration, and flight needs. An additional software module is needed, essentially a virtual Control System Operator. Using defined rules, the Mothership will make decisions concerning the Swarm Members, to the best of its current knowledge. All of this will be documented and downloaded to the Earth-based control center when communications is re-established. An AI capability will be added to Cosmos, in the form of a virtual flight controller agent. Besides the housekeeping functions, we will implement onboard science planning, responsive to on-site conditions, and targets of opportunity.

The Mothership's primary responsibility woll be continuance of the Mission. To a degree, the Cubesats are considered expendable. During communications black-outs, observations will continue, and the Mothership will dispense explorers according to pre-defined rules, and based on it's best on-scene judgment. It will also continue to collect observation science data, and engineering data related to health and performance across the swarm members. Each member of the Swarm is self-documenting. It carries a copy of its Electronic Data Sheet (EDS) description, which will be updated. This defines the system architecture and capabilities, and has both fixed (as-built) and variable entries. The main computer in the Mothership has a copy of all of these, and can get updates by query. The Mothership also has parameters on each unit's state, such as electrical power remaining, temperature, position, etc. One value of the database is, if the Mothership needs a unit with a high resolution imager, it knows what unit that is, and whether it has been deployed or not. If it has been deployed, it will query the unit on its position and health status. Implementing the EDS in a true database has advantages, since the position of the data item in the database also carries information. It also allows the use of off-the-shelf database tools. The individual Cubesats have a "light-weight" version of the database, while the Mothership has a more sophisticated one. All the schema's are the same. The advantage of a formal database is the structure it imposes on the data

There are two parts of the tables, representing static and dynamic data. Static Data represents the hardware and software configuration of the swarm unit. These values are not expected to change during the unit's operation. The Dynamic Data table represents the sensors each particular unit has. These values can change, and the last values will be kept Satellites will exchange two types of data through its communication channel: primary observational data, along with secondary metadata which includes position and localization information along with timing information as a part of the EDS during the mission. This approach was prototyped in a previous project.

The Mothership is responsible for aggregating all of the Cubesats' housekeeping and science data, and transmitting it back to Earth. This is also facilitated by the structure imposed by the database. An Open Source version of an SQL database will be implemented. The EDS documents will be in XML.

Data compression can be implemented onboard the mothership , as well as preliminary data analysis for replanning.

Exploring Comets

There are some 5,253 known comets. The Deep Impact mission returned images of the surface of comet Borrelly in 2001. That surface was hot (26-70C), dry, and dark. In July of 2005, the same mission sent a probe into Comet Tempel-1. It created a crater, allowing imaging of subsurface material. Water ice was seen. Comet Borrely has a coma, which proved to be vaporized subsurface water ice. Deep Impact went on to complete a flyby of Comet Hartley-2 in 2010.

The 1999 Stardust mission retrieved sample material from the tail of Comet Wild 2 and returned it to Earth in 2006. It released a lander, Philae, which successfully touched down on the comet's surface in 2014. The lander communicated with the main spacecraft over a 32kbps link.

Pioneer Venus observed Comet Halley while in transit. This was during a period when the comet was not visible from Earth, because of its proximity to the Sun. The Venus probe monitored the loss of water from the comet as it swung around close to the Sun.

Exploring Mars

Mars, and its two tiny moons and seven Trojans has got some infrastructure in place – a communications relay satellite and a weather satellite. There are several Rovers and landers on the surface. The Viking program was a pair of spacecraft sent to Mars in 1975. Each spacecraft consisted of an orbiter, and a lander. A major target is a Mars sample return mission.

The Mars Pathfinder mission landed on Mars on July 4, 1997. It carried a Rover named Sojourner, which was a 6-wheeled design, with a solar panel for power, but the batteries were not rechargeable. The rest of the lander served as a base station. Communication with the rover was lost in September. It communicated with Earth via the base station using a 9600 baud UHF radio modem. The communication loss leading to end of mission was in the base station communication, while the Rover itself remained functional. The Rover had three cameras, and an x-ray spectrometer.

The MER (Mars Exploration Rovers *Spirit & Opportunity*) are six-wheeled, 400 pound solar-powered robots, launched in 2003 as part of NASA's ongoing Mars Exploration Program. *Opportunity* (MER-B) landed successfully at Meridiani Planum on Mars on January 25, 2004, three weeks after its twin *Spirit* (MER-A) had landed on the other side of the planet. Both used parachutes, a retro-rocket, and a large airbag to land successfully, after transitioning the thin atmosphere of Mars.

The Spirit unit became stuck in 2009, and engineers were unable to free it after 9 months of trying. It was re-tasked as a stationary sensor platform. Contact was lost in 2010.

This is an ongoing mission. It was originally planned for 90 days, but the *Opportunity* Rover is still collecting useful data regarding potential life on our sister planet some 11 years later as of this writing. It has traveled over 35 kilometers on the Martian surface.

The Mars Science Laboratory landed successfully on the Martian surface on August 6, 2012. It had been launched on November 26, 2011. It's location on Mars is the Gale crater. It is designed to operate for two Martian years (sols). The mission is to determine if Mars could have supported life in the past, which is linked to the presence of liquid water.

The Rover vehicle *Curiosity* weights just about 1 ton (2,000 lbs, 900 kg.) and is 10 feet (3 meters) long. It has autonomous navigation, and is expected to cover about 20 km over the life of the mission. The platform uses six wheels

Communication with Earth uses a direct X-band link, and a UHF link to a relay spacecraft in Mars orbit. At landing, the one-way communications time to Earth was 13 minutes, 46 seconds. This varies considerably, with the relative positions of Earth and Mars in their orbits around the Sun.

The science payload includes a series of cameras, including one on a robotic arm, a laser-induced laser spectroscopy instrument, an X-ray spectrometer, and x-ray diffraction/fluorescence instrument, a mass spectrometer, a gas chromatograph, and a laser spectrometer. In addition, the rover hosts a weather station, and radiation detectors. There is cooperation between in-space assets and ground rovers in sighting dust storms by the meteorological satellite in Mars orbit.

NASA's Maven (Mars Atmosphere and Volatile EvolutioN Mission) mission to Mars is an orbiter, to study the Martian atmosphere It will be launched in November 2013, and will reach Mars in September of 2014. It is still operating as of this writing.

The Mars 2020 mission, with its unique helicopter, is currently active on the Martian surface. The rover is focusing on surface geology, and potential biosignatures. It collects samples and leaves them for a future pick up & return mission.It has a core drill, 19 cameras, and a microphone. It recorded and provided the first weather report fromMars.

It touched down in crater Jezero. It is still operating at this writing.The helicopter *Ingenuity*, the first aircraft to operated on another planet is also doing well. As this is being written, the nineth flight was successfully completed.

The Emirates Mars Mission is currently operating, with the Orbiter and the Hope Probe. The mission launched in July of 2020, and entered Mars orbit in February of 2021. It was one of three missions taking advantage of the July 2020 Mars launch window. The others were China and the United States. There have now been five countrys that sent spacecraft to Mars. Hope is about the size of a smal car, and has a mass of a ton and a half. It has two 900 watt solar arrray, and a 1.5 meter high gain antenna. It uses NASA's Deep Space Network for communication.

The Chinese Tianwen mission has an orbiter, a lander, and a rover. The Probe is equipped wiht a multi-band, high resolution imager, and infrared spectrometer, and an ultraviolet spectrometer. The mission consists of 5 parts, an orbiter, deployabe camera, drop camera, a lander, and a Rover. Total mass is 5 tons, and there are 13 instruments. It is focusing on surface geology, Mars internal structureand atmosphere. The lander/rover successfully touched down on the surface in May of 2021.

In 2015, NASA's Small Innovative Missions for Planetary Exploration program got 22 proposals. The Mars Micro Orbiter entered preliminary design review in 2018. It had evolved from a 6U to a 12U size. The winner of the competition was Malin Space Systems of San Diego. The mission was to study the atmosphere, polar caps, energy budget, and characterize habitability.

ExoMars is a program by the European Space Agency and the Russian Space Agency. It has goals in astrobiology, looking for past signs of life on the planet. In 2016, the Trace Gas Orbiter entered Mars orbit. It deployed the Schiaparelli lander to the surface. It unfortunately crashed.

The follow-on ExoMars 2022 mission would launch in September of 2022. The Russians supplied a 1,800 pound lander. It will deploy a rover. It will search for biosignatures.

Marco Project

Mars Cube One (MarCO) is the first interplanetary cubesat mission, headed by JPL. It involves sending two 6U Cubesats to Mars, along with the Insight Rover. The 6U cubesats separated at Earth orbit and proceed on their own. The mission launched in

May, of 2018. The rover is named Perseverance, and the helicopter is Ingenuity.

The Cubesats will serve as a real-time communications relay with Earth during the critical descent and landing phase of the rover. The lander talks to the Cubesat relays over an 8kbps UHF link, and the Cubesats relay this to Earth over an 8kbps X-band link to the DSN.

The Cubesats are stabilized with reaction wheels, and have propulsion systems to unload the wheels, and adjust their orbital position.

In 2016, the NASA's Martian Moons eXploration Community defined a series of mission science objectives. These included determining whether the moon Phobos is a captured asteroid, or the result of an impact. It also defined a goal tio define the origin of the moon Deimos. They also want to study the surface processes on airless small bodies in Mars orbit, gain insight into the Mars surface environment evolution, and gain a better understanding of the atmosphere-ground system and water cycle dynamics. These goals and the mission would be an excellent application for a swarm of Cubesats, similar to the one defined later in this book for Jupiter exploration.

Later in this document, we will discuss the Pinesat Study, a mission based on the Juno mission, but utilizing Cubesats.

Saturn

Saturn and it's 62 known moons has a one-way light time around 1.4 hours. Saturn has been visited by spacecraft four times. The first was a flyby by Pioneer-10 in 1979. This showed the temperature of the planet was 250 degrees K. Voyager-1 visited in 1980. It conducted a close flyby of the moon Titan to study its atmosphere. It is, unfortunately, opaque in visible light. We do know it rains methane. Voyager-2 swung by a year later, and data showed changes in the rings since its sister mission visited the year before. Temperature and pressure profiles of the atmosphere were gathered. Saturn's temperature was measured at 70 degrees above absolute zero at the top of the clouds, and -130 c near the surface. The flybys discovered additional moons, and small gaps in the rings.

Cassini was the fourth spacecraft to study Saturn, which has rings, although smaller than Jupiter. The rings were confirmed by the Voyager spacecraft in the 1980's. Cassini entered into Saturnian orbit, and is still returning data. The one-way communications time varies form 68-84 minutes. It has also collected data on the Saturnian moons Titan, Enceladus, Mimas, Tethys, Dione, Rhea, Iapetus, and Helene. Things are strange in the Saturnian system. Cassini observed a hurricane in 2006 on the planet's south pole. It appears to be stationary, 5,000 miles (8,300 km) across, 40 miles (67 km) high, with winds of 350 mph (560 kph). The large moon Titan has lakes of a liquid hydrocarbon, with possible seas of methane and ethane. Cassini launched a probe *Huygens* to Titan, and it landed on solid ground below the atmosphere. The Cassini mission was responsible for the discovery of seven new moons of Saturn.

Cassini observed a massive storm on Saturn, the great white spot, that recurs every 30 years. The storm, larger than the red one on Jupiter, exhibited a discharge that spiked the temperature 150 degrees. At the same time, Earth observations showed a large increase in atmospheric ethylene gas. It also discovered large lakes or seas of hydrocarbons near the planet's north pole.

Cassini discovered a possible atmosphere on the moon Enceladus, with ionized water vapor, and ice geysers. Many of the Saturnian moons are in tidal lock with their mother planet. Being so close to its giant neighbor Jupiter affects the Saturnian system.

Cassini cost $3 billion ($10^9$) and couldn't get close enough to Saturn's rings because of the ice particles. One hit could cancel the mission. Imagine instead a Cassini-size spacecraft at a safe distance, deploying swarms of hundreds of Cubesats.

Exploring the Ice Giants

The Ice giants are the planets Uranus and Neptune. Their composition differs from that of the gas giants.

Uranus

Uranus has 27 known moons, a 13-ring system, and a one way light time of 2.7 hours from Earth. It has one known Trojan. Uranus was imaged in a flyby by the Voyager-2 spacecraft in 1986. It also captured some images of the Uranian moon Umbriel. But, Uranus and Neptune are one of the great remaining unknowns in the solar system, since neither have been explored in detail, by a dedicated mission. There is a desire to put an explorer spacecraft in orbit, and use that as a platform to launch probes into the atmosphere.

Uranus and Neptune are sometimes referred to as ice giants, since their atmospheres are known to contain water, ammonia, and methane ice. Uranus has a magnetic field. Interestingly, Uranus' spin axis is tilted into the plane of its orbit around the Sun. Seasonal changes and weather have been observed. The Voyager-2 mission imaged Uranus on its way from Jupiter, and out of the solar system. Atmospheric wind speeds are know to approach 900 kilometers per hour. It's orbit period is 84 Earth years. It receives about 1/400th of the light that the Earth does from the Sun, so solar power is probably not a viable choice.

Because of the strange orientation of the planet's rotation axis, during the solstice, one side of the planet faces the Sun continuously, and the other faces deep space. Each pole gets 42 years of direct (though weak) sunlight, and 42 years of darkness. In spite of this, the equator is the hottest region. At this writing, the planet is in its autumnal equinox.

Uranus has a strange predominately water-ammonia ocean, which is electrically conductive. A major targeted mission is the Uranus orbiter and Probe.

Mission analysis comes up with a 12-13 year long cruise from Earth to Uranus.

Neptune

Neptune has 14 known moons, and 18 known Trojans. It's one-way light time is around 4.3 hours. Neptune has also been visited by Voyager-2 in 1989. It discovered six new moons. That is the extent of close-up observations of the planet. Neptune has rings, like Jupiter and Saturn, and a great dark spot. It's moon Triton has geysers and polar caps. Triton has an interesting retrograde orbit – it goes in a different direction than the other moons. Triton's surface is mostly frozen nitrogen, and is geologically active. It is speculated that Triton has a subterranean ocean. The moon Ptoteus is an ellipsoid, not a sphere.

Exploring Pluto, and beyond

Pluto was downgraded from a planet to a Kuiper Belt object. The New Horizons mission to Pluto and the Kuiper Belt began in January of 2006, and reached the vicinity of Pluto in July 2015. It conducted a 6-month survey of Pluto, and went out farther into the Kuiper belt, on an 3 year extended mission, which is ongoing at this writing.

To conserve heat and mass, New Horizon's spacecraft and instrument electronics are housed together in IEMs (Integrated Electronics Modules). There are two redundant IEMs.

In March of 2007, the Command and Data Handling computer experienced an uncorrectable memory error and rebooted itself, causing the spacecraft to go into safe mode. The craft fully recovered within two days, with some data loss on Jupiter's magnetotail. The one-way light time back to Earth is 4.6 hours.

In 2015, the Pluto flyby occurred, and data began to flow back to Earth. It took a year for all the imaging data to be transmitted, due to distances and transmit power involved.

Pluto had one known moon, Charon, before New Horizons Team members, using Hubble Space Telescope data, discovered Nix, Hydra, Styx, and Kerebos.

Kuiper Belt Objects (KBO)

The Kuiper Belt extends from the orbit of Neptune out approximately 50 AU. There are three known dwarf planets, the formal Planet Pluto and two others. Over 100,000 units are speculated to exist. Neptune has a major influence over the Kuiper belt objects. Not much is known about the belt and its objects, since astronomers have had to rely on ground based observation. The New Horizons mission is proceeding out through the Kuiper belt, and will tell us what it sees.

That's just the neighborhood. You want to talk about interstellar Cubesats now? Actually, Cubesats can play a role in Astrophysics and space astronomy by observations of other star systems. In space, particularly behind the Moon, when the Sun's light is blocked, is good good observation point. The L2 point is the destination for the James Webb Space Telescope. An approach that Cubesats can provide is distributed observation, using a station-keeping swarm to implement a large synthetic aperture telescope.

A proposed project to understand the energy transport from black holes would observe in the 5 MHz band. This has to be done away from the Earth's "radio pollution." They postulate a constellation of 100 Cubesats in a 1 kilometer diameter sphere, with 10 cm station-keeping. There is a mothership that serves as the constellation's communication relay, and pre-processes the data to control downlink bandwidth.

Making the Journey – How do we get there?

In 2015, the Planetary Society's LightSail-1 successfully deployed its solar sail. This was done in Earth orbit. Planned follow-on projects LightSail-2, 3, and 4 will follow. Lightsail-2 will be launched shortly after this book goes to press, with 32 square meters of sail, and advanced guidance electronics. It is a 3U Cubesat. It will deploy its sail at 800km. LightSat's 3 and 4 will be more than technology demonstration, with the 4th unit heading to the L1 Lagrange point, to provide earlier warning of Solar geomagnetic storms. Other systems have been proposed with continuous low-thrust ion engines. All these approaches require specific new trajectory designs. There is increasing effort in applying non-linear, non-Keplerian orbits. If you have continuous low-level thrust, you have a non-linear problem in infinite dimensions.

Don't just send one

Here we discuss aggregations of Cubesats. They may interact with each other, or not. Some of the architectures include trains, Constellations, and clusters.

Clusters

In computing, a cluster means a group of loosely coupled elements, working together on the same problem. If the cluster were more tightly coupled, and self-directed, we'd have a swarm. As it is, we would have a group of individuals that could be considered a single entity. Generally, members of a cluster have the same hardware and software configuration. One issue in clustering is the degree of coupling between elements. No coupling means we have a mob. A lot of coupling and we might have a swarm. In a cluster, management of the cluster can be centralized in a "boss" or can be distributed.

Swarms

This section describes a different approach: collections of smaller co-operating systems that can combine their efforts and work as ad-hoc teams on problems of interest. Cubesats can be organized in Swarms.

This is based on the collective or parallel behavior of homogeneous systems. This covers collective behavior, modeled on biological systems. Examples in nature include migrating birds, schooling fish, and herding sheep. A collective behavior emerges form interactions between members of the swarm, and the environment. The resources of the swarm can be organized dynamically.

Exploring the asteroids requires a diverse and agile system. Thus, a swarm of small spacecraft with different capabilities might be used, combining into Teams of Convenience to address situations and issues discovered in situ.

In Swarm robots, the key issues are communication between units, and cooperative behavior. The capability of individual units odes not much matter; it is the strength in numbers. Ants and other social insects such as termites, wasps, and bees, are models for robot swarm behavior. Self-organizing behavior emerges from decentralized systems that interact with members of the group, and the environment. Swarm intelligence is an emerging field, and swarm robotics is in its infancy.

For a constellation of Cubesats, the Swarm behavior of peer units could be implemented.

An actual "constellation" of fifty 2U and 3U Cubesats were deployed in 2015. Some were released from the ISS, and some from a rocket launch. They collected and telemetered data on the lower thermospere. This is not a Constellation, per se, but 50 units acting on their own, reporting back to their home institutions. Universities around the world participated, and built units from the QB50 specification.

The data comes from the region below 85 kilometers, which has enough of an atmosphere to impede spacecraft. The Cubesats collect data as long as they can, while they are reentering the atmosphere. At these altitudes, the rarefied atmosphere can reach 2,500 degrees C. It is also a region where the dynamics are controlled by atmospheric tides, themselves controlled by diurnal heating and cooling. The member Cubesats do not interact with each other.

Case Study – Swarms of Cubesats to Jupiter

The Pinesat was a strawman mission, used as a teaching tool by the author in a Cubesat Engineering and Operations course in the Summer of 2016. The parameters of the Juno mission were used to bound the problem, and examine the possibility of replacing one large exploration spacecraft with a number of smaller ones.

The Cubesat alternative

The author and a group of highly motivated and skilled students conducted this project. We used the parameters of the Juno mission as far as possible to define the mission. The following items and subsystems will be copied and reused from the Juno mission, and are at TRL-9:

Solar panels, power distribution unit, Flight Computers (RAD-750), batteries, X-band transceiver and high gain antenna, bi-propellant LEROS1b main engine for trajectory correction, mono-propellant attitude control thrusters.

The Atlas-V-551 launch vehicle, TRL-9, many launches to date, including one to Jupiter.

For the proposed new parts, we assessed the technology, and assigned these TRL levels. It has to be taken into account that a TRL derived for an Earth mission might need to be adjusted downward for a long interplanetary cruise, and for operation at Jupiter.

Although Cubesats themselves are at TRL-7, the concept of cubesat Swarms and Clusters are at TRL levels 3-4.

The mechanical design of Pinesat was straightforward. The name comes from the fact that the dispenser resembles a pine tree. The dispenser has a central hexagonal tube, with the propulsion and electrical power section at the launch vehicle interface end. The avionics and data storage are located in the nose of the vehicle while p-pod's are distributed radially along the central tube. This allows for longitudinal deployment. We postulate having 333 p-pod class dispensers, each with a 3-U cubesat inside.

The mothership/dispenser will have a bi-propellant engine for orbit and cruise adjustments, and a mono-propellant system for attitude control and reaction wheel momentum unloading.

One advantage of the Pinesat is, like the Shuttle, payloads can be tested before deployment from the carrier. Known bad units can be discarded into Jupiter's atmosphere. The carrier is designed to be modular and adaptable. It is scaleable to 100's or 1,000's of Cubesats.

The Pinesat will keep a database of Electronic Data Sheets of all the Cubesats. An electronic data sheet is a spec sheet, just not in printed form. It is, in essence, a database entry, containing all of the parameters of the system. This includes data like state-of-charge, capabilities, processor speed, memory capacity, operational status, instrument complement, what ever is applicable. This can be updated by a query request from the dispenser's main computer. The dispenser/mothership's main computer will also store its own vehicle's EDS. These structures are for communications and storage of technical data, in a defined format. Here is a simplified EDS for a Cubesat's Raspberry Pi flight computer.

Electronic Data sheet - Cluster Cubesat				6/13/2016
Static				
item	value	description	units	
1	4	cpu cores		Main cpu
2	A7	arch		
3	900	clock	MHz	
4	tbd	flash	mbytes	
5	512	sram	kbytes	
6	17	dig I/O	lines	
7	0	analog in	lines	
8	0	analog out	lines	
9	4	serial I/O	lines	
10	8.4 x 5.7	size	cmxcm	
11	66	weight	grams	
12	Debian	opsys		
13	5	supply	volt	

The Pinesat dispenser will use the same dual redundant, rad-hard RAD-750 flight computers used on Juno. They have 256 megabytes of flash, 128 megabytes of DRAM, and operate at 200 MHz. They will be running Linux. They will have sufficient storage for the database of the Cubesats' EDS. They will also host an onboard network for communications with the Cubesats when they are onboard, and via radio when they have been deployed. The dispenser vehicle hosts the Swarm's communications links with Earth.

Cubesats

The Cubesats will be 3U format, with identical busses, and varying science instrument payloads. They will have a sun sensor and a magnetic field sensor. Magnetic torquers may be included. They may include a cold gas propulsion system.

The Cubesats will use a Raspberry-Pi flight computer, running NASA/GSFC's CFS/CFE flight software. They will have a sun sensor and a magnetic field sensor.. For the payload, different instrumentation will be included on different Cubesats. The Cubesats will be deployed by the mothership as required, to observe and collect data on targets of interest. The Swarm's instrument complement will be defined by the Mission scientists, partially determined by what the Juno mission uncovers.

Unlike Earth Cubesat missions, the Cubesats going to Jupiter can have their own propulsion. The big limiting factor for them is electrical power. They can't carry solar arrays large enough to make sense. They will be dispersed from the carrier fully charged, and operate as long as they can. The electronics and software will be optimized to minimize power usage.

Ops concept

The Mothership transports the Cubesats to Jupiter unpowered. Every day or week (tbd), the units are powered on, one at a time, and checked for functionality. The onboard database is updated as required. The results are sent back to the control center on Earth.

Most of the mission goals are preplanned and stored in the mothership's computers, but the spacecraft also will have the flexibility to respond to unanticipated events.

After orbital insertion at Jupiter, and another system check of itself and the Cubesats, the Mothership will deploy a series of Cubesat scouts on a reconnaissance missions, to seek out areas of interest. The Mothership first deploys Cubesats with broad spectral sensing. Based on their findings, the mothership will deploy Cubesats with more specific instrumentation to the area of interest. (For example, an advanced thermal imager to an area of observed thermal activity). The Cubesats are released in the order of necessity. The Cubesats will be able to adjust their attitude with torquer bars to push against the large Jovian magnetic field. We can also include the idea of Cubesat "suicide mission," where one Cubesat plunges into the ring system, or Jupiter's upper atmosphere, and returns data until it is rendered non-functional. The entry and impact, if any, can be observed by other Cubesats.

The Cubesats would:

- Conduct radio occultation experiments to better categorize the distribution of particles in the rings, by size.

- Explore the characteristics of the ring systems, including density, size, distribution, and particle composition. This can also be conducted with synchronized simultaneous observation from multiple observation points.

- Explore features visible on on the planetary "surface."

- Map the Jovian magnetic field and trapped charged particle environment.

- Be able to respond to targets of opportunity, should they arise, such as the observed plunge of comet Shumaker-Levy-3 into Jupiter's atmosphere by the Galileo spacecraft.

The Cubesat instruments will be expanded beyond that of Juno. Standard Cubesat busses will support different sensor suites. We have, essentially, 334 observation platforms, including the dispenser. This provides greater mission flexibility.

Enabling Technologies

This section defines the technologies that will be used to implement. A subsequent section will examine their technology readiness levels (TRL's).

Better batteries and solar panels

The batteries and solar panels on the Juno mission are functioning well, and since the Mothership will be roughly the same size, we could postulate their use. A design study will be needed to see if a steerable array is needed. Due to technology advances, solar cells can be used out to 5 AU. In 2012, the best efficiency was under 30%. That has improved a bit since then. Solar cells also degrade from exposure to radiation.

Rad-Hard Software

This is a concept that implements routines that check and self-check, report, and attempt to re-mediate. It is an outgrowth of the testing and self-testing of the computers' functionality, with focus on detection of radiation induced damage. We know, for example, that one of the tell-tales for radiation damage is increasing current draw. At the same time, we monitor other activities and parameters in the system. This partially addresses the problem of operating with non-radiation hardened hardware in a high radiation environment.

From formal testing results, and with certain key engineering tools, we can come up with likely failure modes, and possible remediations. Besides self-test, we can have cross-checking of systems. Not everything can be tested by the software, without some additional hardware. First we will discuss the engineering analysis that will help us define the possible hardware and software failure cases, and then we will discuss possible actions and remediations. None is this is new, but the suggestion is to collect together best practices in the software testing area, develop a library of RHS routines, and get operational experience. Another advantage of the software approach is that we can change it after launch, as more is learned, and conditions change.

Rad Hard software is a series of software routines that run in the background on the flight computer, and check for the signs of radiation damage. The biggest indicator is an increase in current draw. The flight cpu must monitor and trend it's current draw, and take critical action such as a reboot if it deems necessary. The Rad Hard software is a variation on self-check routines, but with the ability to take action if needed. It can keep tabs on memory by conducting CRC (cyclic redundancy checks), and one approach to mitigating damage to semiconductor memory is "scrubbing," where we read and write back each memory locations (being careful not to interfere with ongoing operations). This can be done by a background task that is the lowest priority in the system. Watchdog timers are also useful in getting out of a situation such as the Priority Inversion, or just a radiation-induced bit flip. There should be a pre-defined safe mode for the computer as well. Key state data from just before the fault should be stored and telemetered to the control center. Unused portions of memory can be filled with bit patterns that can be monitored for changes. We must be certain that all of the unused interrupt vectors point to a safe area in the code.

Functions within the RHS include current monitoring as a tell-tale of radiation damage, self-diagnosis suite, spurious interrupt test, memory test(s), checksum over code, data corruption testing, memory scrub, I/O functionality test, peripherals test, stack overflow monitoring, and a watchdog timer. A complete failure modes and effects analysis will be conducted over the flight computer and associated sensors and mechanisms, and this will be used to scope the RHS. The systems will keep and report trending data on the flight electronics. In most cases, the only remediation will be a reboot. However, since the units will have identical configurations, the data will be useful to be able to predict pending failures, and to possible correct them.

Dispenser/Mothership

The mothership will be built with standard aerospace products with a Jupiter-mission heritage. We expect to be able to use the same batteries and solar panels from the Juno mission, since the Mothership will be roughly the same size. A design study will be needed to see if a steerable array is needed. The X-band transceiver on Juno would be a candidate for the Earth link. Standard p-pod cubesat dispensers are baselined, but the affects of long storage of the CubeSats in the dispensers must be considered. The effects of cold welding during the 5 year transit needs to be studied.

Deployment of Cubesats from the p-pod after 5 years exposed to the space environment has to be examined. The problem is that the Cubesat may not deploy, due to cold-welding of it to the dispenser. Cold welding occurs with similar materials touching in a vacuum.

Inter-unit communications

Before deployment, the Cubesat flight computers use the network infrastructure of the Mothership to communicate. After deployment, a radio solution may be applied. It is possible that the Cubesats will use the mothership as their communication relay, and not implement cubesat-to-cubesat direct communications. Cubesat to Mothership could be S-band. It is not necessarily to implement a delay tolerant protocol, since the Cubesats will be "in the vicinity" of the Mothership.

Compute cluster of convenience – Beowulf

Using a variation of the Beowulf clustering software and the intranet of the Mothership, the Cubesats awaiting deployment can be linked into a Compute cluster configuration. Each compute node will have the Beowulf software pre-loaded as part of its Linux operating system.

Beowulf was developed to provide a low cost solution to linking commodity pc's into a supercomputer. The approach has been applied to clusters of small architectures such as those that serve as flight computers for Cubesats. As part of the student summer program, we demonstrated a 5 node cluster, using Raspberry Pi's. Several 64-node clusters have been constructed.

The Beowulf cluster is ideal for sorting and classifying data; an example application is the Probabilistic Neural Network. This algorithm has been used to search for patterns in remotely sensed data. It is computationally intensive, but scales well across compute clusters. It was developed by the Adaptive Scientific Data Processing (ASDP) group at NASA/GSFC. The program is available in Java source code.

The first Beowulf cluster to be flown in space was built from 20 206-MHz StrongARM (SA1110) processors, and flew on the X-Sat, which was Singapore's first satellite. The performance was 4,000 MIPS. And the cluster drew 25 watts. The satellite was a 100 kg unit, 80 CM cube. The cluster was used because the satellite collected large amounts of image data (80 GB per day), most of which was not relevant to the mission. An onboard classification algorithm selected which images would be downloaded. For example, cloudy images were discarded, since land images were of interest.

Wrap-up

Cubesats have proven themselves in Earth orbit. About half of the Cubesat missions fail because of lack of experience by schools or individuals. When NASA builds a Cubesat, it goes through the same rigorous design process at any other spacecraft mission they do. Cubesats are a game changer, because they allow a lot more science, for a lot less cost. The first "Interplanetary" Interplanetary Cubesats have been launched. Cubesat swarms provide economies of scale to drive mission costs down, and mission data capture up. There are massive challenges in space, once we get away from the Earth. But, before 1957, there were massive challenges in getting to orbit, and in 1969, massive challenges getting men to the Moon. All this was solved with good engineering, and persistence. The United States is the first country whose spacecraft have visited all the planets. It will soon be the first nation whose Cubesats have visited a planet other than Earth.

I really hope to be able to name the sequel to this book, "Interstellar Cubesats."

References

Alvarez, Jennifer L.; Rice, John R. Samson, Jr., Michael A. Koets. "Increasing the Capability of Cubesat-based Software-Defined Radio Applications," avail: ieeexplore.ieee.org/document/7500847/

Ardila, David R. "Cubesats for Astrophysics," Aerospace Corp. avail: https://cor.gsfc.nasa.gov/copag/rfi/CubesatsforAstrophysics.pdf

Baisamo, James M. et al "CubeSat technology adaption for in-situ characterization of NEOs," presentation, avail: NASA Technical Reports Server (NTRS), 2014, document id 20140004799.

Betancourt, Mark "CubeSats to the Moon (Mars and Saturn, too)", Air & Space Magazine, Sept 2014.

Budianu, A. et al.. "Inter-satellite links for Cubesats". In: IEEE Aerospace Conference Proceeding, 2013, pp. 1–10. avail: ieeexplore.ieee.org/document/6496947/

Challa, Obulapathi N., McNair, Janise "CubeSat Torrent: Torrent-like distributed communications for CubeSat satellite clusters," Military Communications Conference, 2012 , ((MILCOM 2012) July 19, 2016

Challa, Obulapathi N., McNair, Janise "Distributed Data Storage on Cubesat Clusters, Advances in Computing 2013, (3)3 pp.36-49 Electronics and Computer Engineering, U. Florida, Gainsville.

Clark, P. E. et al "BEES for ANTS: Space Mission Applications for the Autonomous NanoTechnology Swarm," AIAA, Sept. 2004. avail: https://www.researchgate.net/publication/265158396_BEES_for_

ANTS_Space_Mission_Applications_for_the_Autonomous_Nano Technology_Swarm.

Copernicus, Nicolaus *On the Revolutions of the Heavenly Spheres*, 1543, ASIN-B01MS8TGOV.

Cudmore, Alan Pi-Sat: A Low Cost Small Satellite and Distributed Mission Test Program, NASA/GSFC Code 582, avail: https://ntrs.nasa.gov/archive/nasa/casi.ntrs.nasa.gov/20150023353.pdf

Curtis, S. A. et al "Use of Swarm Intelligence in Spacecraft Constellations for the Resource Exploration of the Asteroid Belt," 2003, Third International Workshop on Satellite Constellations and Formation flying, Pisa, Italy.

Fortescue, Peter and Stark, John *Spacecraft System Engineering*, 2nd ed, Wiley, 1995, ISBN 0-471-95220-6.

Gilster, Paul "CubeSats: Deep Space Possibilities," Sept. 2015, avail: http://www.centauri-dreams.org/?p=34056.

Johnson, Les et al "Solar Sail Propulsion for Interplanetary Cubesats," NASA/MSFC.

Kreck Institute of Space Studies, *Small Satellites: A revolution in Space Science*, Final Report, July 2014. avail: kiss.caltech.edu/study/smallsat/KISS-SmallSat-FinalReport.pdf

Lappas, Vaios, et al "CubeSail: A low cost Cubesat based solar sail demonstration mission," Advances in Space Research, 2001, 48.11, 1890-1901.

Macdonald, Malcolm "Advances in Solar Sailing,"2014, Springer Praxis, ISBN-978-3642349065.

NASA, "Hitchhiking Into the Solar System: Launching NASA's First Deep-Space Cubesats," avail: www.nasa.gov/exploration.

Ross, Shane, "The Interplanetary Transport Network," American Scientist, Vol 94, May-June 2006

Spangelo, et al "JPL's Advanced CubeSat Concepts for Interplanetary Science and Exploration Missions, Cubesat Workshop," 2015, California Institute of Technology, JPL. Avail; https://digitalcommons.usu.edu/cgi/viewcontent.cgi?article=3313&context=smallsat

Staehle, Robert et al, "Interplanetary Cubesats: Opening the Solar System to a Broad Community at Lower Cost" Cubesat Workshop, 2011, Logan Utah. Avail: https://www.nasa.gov/pdf/716078main_Staehle_2011_PhI_Cubesat.pdf

Stakem, Patrick H. "Free Software in Space–the NASA Case," invited paper, Software Livre 2002, May 3, 2002, Porto Allegre, Brazil.

Stakem, Patrick H.; Rezende, Aryadne; Ravazzi, Andre "Cubesat Swarm Communications," 2016.

Stakem, Patrick H.; Da Costa, Rodrigo Santos Valente; Rezende, Aryadne; Ravazzi, Andre "A Cubesat-based alternative for the Juno Mission to Jupiter, 2017, available from the author, pstakem@gmail.com

Stakem, Patrick H., Kerber, Jonathas "Rad-hard software, Cubesat Flight Computer Self-monitoring, Testing, Diagnosis, and Remediation," 2017, available from the author, pstakem1@jhu.edu.

Stakem, Patrick H. "Lunar and Planetary Cubesat Missions," March Volume 15, Polytech Revista de Tecnologia e Ciência, avail: http://www.polyteck.com.br/revista_online/ed_15.pdf

Tan, Ying *GPU-based Parallel Implementation of Swarm Intelligence Algorithms*, 2016, 1st ed, Morgan Kaufmann, ISBN-978-0128093627.

Truszkowski, Walt *Autonomous and Autonomic Systems: With Applications to NASA Intelligent Spacecraft Operations and Exploration Systems*, Springer; 1st Edition. edition, 2009, ISBN-1846282322.

Truszkowski, Walt, et al. *ANTS: Exploring the Solar System with an Autonomous Nanotechnology Swarm*. J. Lunar and Planetary Science XXXIII (2002)

Virgili, Bastide et al "Mega-constellations Issues," 41st COSPAR Scientific Assembly, 2016, avail: http://cospar2016.tubitak.gov.tr/en/

Resources

- Small Spacecraft Technology State of the Art, NASA-Ames, NASA/TP2014-216648/REV1, July 2014.
- Core Flight System (CFS) Deployment Guide, Ver. 2.8, 9/30/2010, NASA/GSFC 582-2008-012.
- "NOAA Space Weather Scale for Radio Blackouts". NOAA / Space Weather Prediction Center. 2005-03-01.
- Cubesat Design Specification, Cubesat Program, California Polytechnic State University, avail.https://www.google.com/searchq=Cubesat+Design+Specification&ie=utf-8&oe=utf-8
- www.icubesat.org, Interplanetary Cubesat Workshop.
- https://www.nasa.gov/feature/goddard/2017/nasa-studies-cubesat-mission-to-solve-venusian-mystery

- Cubesat Concept and the Provision of Deployer Services, avail:https://eoportal.org/web/eoportal/satellite-missions/content/-/article/Cubesat-concept-1
- NASA Systems Engineering Handbook, NASA SP-2007-6105. Avail:https://ntrs.nasa.gov/archive/nasa/casi.ntrs.nasa.gov/20080008301.pdf
  - http://sen.com/news/cubesats-set-for-new-role-as-planetary-explorers
  - https://solarsystem.nasa.gov
  - http://ssed.gsfc.nasa.gov/pcsi
  - New Horizons Mission - https://www.nasa.gov/mission_pages/newhorizons/overview/index.html
  - www.planetary.org
  - Visions and Voyages for Planetary Science in the Decade 2013 – 2022 (2011). avail:
  - https://www.nap.edu/catalog/13117/vision-and-voyages-for-planetary-science-in-the-decade-2013-2022.
  - JPL Small-body Database. Avail:https://ssd.jpl.nasa.gov/sbdb.cgi
  - https://pds-rings.seti.org/jupiter/
  - http://astronomy.swin.edu.au/cosmos/C/Centaurs
  - https://svs.gsfc.nasa.gov/
  - https://voyager.jpl.nasa.gov/mission/science/
  - http://www.iflscience.com/space/how-saturns-shepherd-moons-herd-its-rings/
  - http://www.airspacemag.com/space/cubesats-moon-mars-and-saturn-too-180952389/
  - http://phys.org/news/2015-11-cubesats-deep-space.html
  - https://www.jpl.nasa.gov/missions/interplanetary-nano-spacecraft-pathfinder-in-relevant-environment-inspire
  - Wikipedia, various.

Glossary

1U – one unit for a Cubesat, 10 x 10 x 10 cm.
3U – three units for a Cubesat
6U – 6 units in size, where 1u is defined by dimensions and weight.
802.11 – a radio frequency wireless data communications standard.
AACS – (JPL) Attitude and articulation control system.
ACE – attitude control electronics
AFB – Air Force Base.
AGC – Automated guidance and control.
AIAA – American Institute of Aeronautics and Astronautics.
AIST – NASA GSFC Advanced Information System Technology .
ALU – arithmetic logic unit.
ANSI – American National Standards Institute
ANTS – autonomous nano technology swarm
Antares – Space launch vehicle, compatible with Cubesats, by Orbital/ATK (U.S.)
AP – application programs.
API – application program interface; specification for software modules to communicate.
APL – Applied Physics Laboratory, of the Johns Hopkins University.
Apm – antenna pointing mechanism
Apollo – US manned lunar program.
Arduino – a small, inexpensive microcontroller architecture.
ASIC – application specific integrated circuit
ASIN – Amazon Standard Inventory Number.
Asteroid - minor planets, orbiting the Sun.
async – non synchronized
ATP – authority to proceed
AU – astronomical unit. Roughly 149.6 million kilometers, the mean distance between Earth and Sun.
BAE – British Aerospace.
Baud – symbol rate; may or may not be the same as bit rate.

Beowolf – a cluster of commodity computers; multiprocessor, using Linux.
Binary – using base 2 arithmetic for number representation.
BIST – built-in self test.
Bit – binary variable, value of 1 or 0.
Bow shock- Where thr solar wind begins to interact with a planet's magnetosphere.
Centaur – a minor planet in an unstable orbit, behaving like an asteroid or comet.
Comet – icy body orbiting the Sun inn a very eccentric orbit.
BP - bundle protocol, for dealing with errors and disconnects.
BSP – board support package. Customization Software and device drivers.
Buffer – a temporary holding location for data.
Bug – an error in a program or device.
Bus – an electrical connection between 2 or more units; the engineering part of the spacecraft.
Byte – a collection of 8 bits,
CalPoly – California Polytechnic State University,. San Luis Obispo, CA.
CAN - controller area network bus.
CCSDS – Consultive Committee on Space Data Systems.
CDR – critical design review
C&DH – Command and Data Handling
CDFP CCSDS File Delivery Protocol
Centaur – a minor planet in an unstable orbit, behaving like an asteroid or comet.
Comet – icy body orbiting the Sun inn a very eccentric orbit.
cFE – Core Flight Executive – NASA GSFC reusable flight software.
CFS – Core Flight System – NASA GSFC reusable flight software.
Chip – integrated circuit component.
Clock – periodic timing signal to control and synchronize operations.
CME – Coronal Mass Ejection. Solar storm. .

CogE – cognizant engineer for a particular discipline; go-to guy; specialist.
Comet – icy body orbiting the Sun inn a very eccentric orbit.
Constellation – a grouping of satellites.
COP – computer operating properly.
COTS – commercial, off the shelf
CPU – central processing unit
CRC – cyclic redundancy code – error detection and correction mechanism.
Cubesat – small inexpensive satellite for colleges, high schools, and individuals.
Daemon – in multitasking, a program that runs in the background.
DARPA – (U. S.) Defense advanced research projects agency.
Dnepr – Russian space launch system compatible with Cubesats.
DOD – (U. S.) Department of Defense.
DOE – (U. S.) Department of Energy.
DOF – degrees of freedom.
Downlink – from space to earth.
DSN – Deep Space Network
DSP – digital signal processing/processor.
DTE – Direct-to-Earth
DTN – delay tolerant network; disruption tolerant network.
DUT – device under test.
ECC – error correcting code
Ecliptic – apparent path of the Sun throughout the year.
EDAC – error detection and correction.
EDL – entry, descent, and landing.
EDS – Electronic Data Sheets
EGSE – electrical ground support equipment
EIA – Electronics Industry Association.
ELV – expendable launch vehicle.
Embedded system – a computer systems with limited human interfaces and performing specific tasks. Usually part of a larger system.
EMC – electromagnetic compatibility.
EMI – electromagnetic interference.

EOL – end of life.
EOS – Earth Observation spacecraft.
Ephemeris – orbital position data.
EPS – electrical power subsystem.
ESA – European Space Organization.
ESRO – European Space Research Organization
ESTO – NASA/GSFC – Earth Science Technology Office.
Ethernet – networking protocol, IEEE 802.3
EU – European Union
ev – electron volt, unit of energy
EVA – extra-vehicular activity.
Exception – interrupt due to internal events, such as overflow.
EXPRESS racks – on the ISS, EXpedite the PRocessing of Experiments for Space Station Racks
FAA – (U S.) Federal Aviation Administration.
Fail-safe – a system designed to do no harm in the event of failure.
Falcon – launch vehicle from SpaceX.
FCC – (U.S.) Federal Communications Commission.
FDC – fault detection and correction.
Firewire – IEEE-1394 standard for serial communication.
Flag – a binary state variable.
Flash – non-volatile memory
Flatsat – prototyping and test setup, laid out on a bench for easy access.
FlightLinux – NASA Research Program for Open Source code in space.
Floating point – computer numeric format for real numbers; has significant digits and an exponent.
FPGA – field programmable gate array.
FPU – floating point unit, an ALU for floating point numbers.
Full duplex – communication in both directions simultaneously.
FRAM – ferromagnetic RAM; a non-volatile memory technology
FRR – Flight Readiness Review
FSW – flight software.
FTP – file transfer protocol
Gbyte – 10^9 bytes.

GEO – geosynchronous orbit.
GeV – Giga (10^9) electron volts.
Giga - 10^9
GNC – guidance, navigation, and control.
Gnu – recursive acronym, gnu is not unix.
GPIO – general purpose I/O.
GPL – gnu public license used for free software; referred to as the "copyleft."
GPS – Global Positioning system – Navigation satellites.
GPU – graphics processing unit. ALU for graphics data.
GSFC – Goddard Space Flight Center, Greenbelt, MD.
Gyro – (gyroscope) a sensor to measure rotation.
Half-duplex – communications in two directions, but not simultaneously.
Handshake – co-ordination mechanism.
HDL – hardware description language
Heliophysics – physics of the Sun.
Hertz – cycles per second.
Hi-rel – high reliability
I^2C – a serial communications protocol.
IARU – International Amateur Radio Union
IAU – International Astronomical Union
ICD – interface control document.
IC&DH – Instrument Command & Data Handling.
Ice giant – A large icy/liquid planet, consisting of elements heavier than hydrogen and helium.
IEEE – Institute of Electrical and Electronic engineers
IEEE-754 – standard for floating point representation and calculation.
IIC – inter-integrated circuit (I/O).
IMU – inertial measurement unit.
Integer – the natural numbers, zero, and the negatives of the natural numbers.
Interrupt – an asynchronous event to signal a need for attention (example: the phone rings).
IP – intellectual property; Internet protocol.

IP core – IP describing a chip design that can be licensed to be used in an FPGA or ASIC.
IP-in-Space – Internet Protocol in Space.
IR – infrared, 1-400 terahertz. Perceived as heat.
IRAD – Independent Research & Development.
ISA – instruction set architecture, the software description of the computer.
ISBN – International Standard Book Number.
ISO – International Standards Organization.
ISR – interrupt service routine, a subroutine that handles a particular interrupt event.
ISS – International Space Station
I&T – integration & test
ITAR – International Trafficking in Arms Regulations (US Dept. of State)
ITU – International Telecommunications Union
IV&V – Independent validation and verification.
JEM – Japanese Experiment Module, on the ISS.
JHU – Johns Hopkins University.
JOI – Jovian orbit insertion,
Jovian – pertaining to Jupiter.
JPL – Jet Propulsion Laboratory
JSC – Johnson Space Center, Houston, Texas.
JTAG – Joint Test Action Group; industry group that lead to IEEE 1149.1, Standard Test Access Port and Boundary-Scan Architecture.
JWST – James Webb Space Telescope – follow on to Hubble.
KBO – Kuiper belt object.
Kbps – kilo (10^3) bits per second.
KBO – Kuiper Belt Object
Kg – kilogram.
kHz – kilo (10^3) hertz
Kuiper Belt – beyond Neptune, a ring of small icy asteroids and minor planets.
Ku band – 12-18 Ghz radio

Lagrange (L) point – a null point in the gravity field in the 3-body program.
> L1 - the Lagrange point between the 2 bodies.
> L2 – the Lagrange point behind the smaller body.
> L3 – the Lagrange point behind the larger body.
> L4- the leading Lagrange in an orbit.
> L5 – the trailing Lagrange point in an orbit.

Lan – local area network, wired or wireless.
LaRC – (NASA) Langley Research Center.
Latchup – condition in which a semiconductor device is stuck in one state.
Lbf – pounds-force (0.7 newton-meter)
LEO – low Earth orbit.
Let- Linear Energy Transfer
LGM – little green men.
Lidar – optical radar.
Linux – open source operating system
LRR – launch readiness review
LRU – least recently used; an algorithm for item replacement in a cache.
LSB – least significant bit or byte.
LSP – (NASA) launch services program, or launch services provider
LUT – look up table.
Magnetosphere – a space surrounding a planet or moon that is affected by the primary's magentic field.
Magentopause – abrupt boundary between a magnetosphere and the solar wind.
Magnetotail – magnetosphere extends away from the planet and the Sun.
Master-slave – control process with one element in charge. Master status may be exchanged among elements.
Mbps – mega (10^6) bits per second.
Mbyte – one million (10^6 or 2^{20}) bytes.
Memory leak – when a program uses memory resources but does not return them, leading to a lack of available memory.

Memory scrubbing – detecting and correcting bit errors.
MEMS – Micro Electronic Mechanical System.
MESI – modified, exclusive, shared, invalid state of a cache coherency protocol.
MEV – million electron volts.
MHz – one million (10^6) Hertz
Microcontroller – monolithic cpu + memory + I/O.
Microkernel – operating system which is not monolithic, functions execute in user space.
Microprocessor – monolithic cpu.
Microsat – satellite with a mass between 10 and 100 kg.
Microsecond – 10-6 second.
MJS-77 circa 1977 mission to Mars, Jupiter, Saturn. Name changed to Voyager.
MLI – multi-layer insulation.
MPA – multiple payload adapter for deploying multiple p-pod's
MPE – Maximum predicted environments.
Mph – miles per hour
mram – magnetorestrictive random access memory.
mSec – Millisecond; (10^{-3}) second.
MIPS – millions of instructions per second.
MMU – memory management unit; manned maneuvering unit.
MSB – most significant bit or byte.
Multiplex – combining signals on a communication channel by sampling.
Multicore – multiple processing cores on one substrate or chip; need not be identical.
Mutex – a software mechanism to provide mutual exclusion between tasks.
Nano – 10^{-9}
NanoRacks – a company providing a facility onboard the ISS to support Cubesats
nanoSat – small satellite with a mass between 1 and 10 kg.
NASA - National Aeronautics and Space Administration.
NDA – non-disclosure agreement; legal agreement protecting IP.
NEA – near Earth asteroid

NEC – near Earth Comet
NEN – (NASA's) Near Earth Network
NEO – near Earth object.
Nibble – 4 bits, ½ byte.
NIST – (U.S.) National Institute of Standards and Technology, previously, National Bureau of Standards.
NMI – non-maskable interrupt; cannot be ignored by the software.
NOAA – (U.S.) National Oceanographic and Atmospheric Administration.
Normalized number – in the proper format for floating point representation.
NRCSD - NanoRack CubeSat Deployer
NRE – non-recurring engineering; one-time costs for a project.
NSF – (U.S.) National Science Foundation.
NSR – non-space rated.
NTIA (U.S.) National Telecommunications and Information Administration
NUMA – non-uniform memory access for multiprocessors; local and global memory access protocol.
NVM – non-volatile memory.
NWS – (U.S.) National Weather Service
Nyquist rate – in communications, the minimum sampling rate, equal to twice the highest frequency in the signal.
OBC – on board computer
OBD – On-Board diagnostics.
OBP – On Board Processor
Off-the-shelf – commercially available; not custom.
One-way light time – a measure of distance, in terms of how long it would take light to travel the distance.
Orbit – the path of one body around another, that are linked by gravity.
OpAmp – operational amplifier; linear gain and isolation stage.
OpCode – encoded computer instruction.
Open source – methodology for hardware or software development with free distribution and access.

Operating system – software that controls the allocation of resources in a computer.
OSAL – operating system abstraction layer.
OSI – Open systems interconnect model for networking, from ISO.
Overflow - the result of an arithmetic operation exceeds the capacity of the destination.
Packet – a small container; a block of data on a network.
Paging – memory management technique using fixed size memory blocks.
Paradigm – a pattern or model
Paradigm shift – a change from one paradigm to another. Disruptive or evolutionary.
Parallel – multiple operations or communication proceeding simultaneously.
Parity – a simple error detecting mechanism involving an extra check bit in the word.
PC-104 – standard for a board (90 x 96 mm), a nd a bus for embedded use.
PCB – printed circuit board.
PCSI – (NASA/GSFC) Planetary Cubesat Science Institute
pci – personal computer interface (bus).
PCM – pulse code modulation.
PCSI – NASA's Planetary Cubesat Science Institute
PDCO – NASA's Planetary Defense Coordination Office
PDR – preliminary design review
Perhelion – in an orbit, the closest point to the Sun.
Peta - 10^{15} or 2^{50}
PHO – potentially hazardous object
Phonesat – small satellite using a cell phone for onboard control and computation.
Picosat – small satellite with a mass between 0.1 and 1 kg.
Piezo – production of electricity by mechanical stress.
Pinout – mapping of signals to I/O pins of a device.
Pipeline – operations in serial, assembly-line fashion.
PiSat – a Cubesat architecture developed at NASA-GSFC, based on the Raspberry Pi architecture.

Pixel – picture element; smallest addressable element on a display or a sensor.
PLL – phase locked loop.
PocketQube – smaller than a Cubesat; 5 cm cubed, a mass of no more than 180 grams, and uses COTS components.
Poc – point of contact
POSIX – IEEE standard operating system.
PPF – payload processing facility
PPL – preferred parts list (NASA).
P-POD – Cubesat launch dispenser, Poly-Picosatellite Orbital Deployer
Psia – pounds per square inch, absolute.
PSP – Platform Support Package.
Rad – unit of radiation exposure
Rad750 – A radiation hardened IBM PowerPC cpu.
Ram – random access memory.
RBF – remove before flight.
Real-time – system that responds to events in a predictable, bounded time.
Reset – signal and process that returns the hardware to a known, defined state.
RF – radio frequency
RFC – request for comment
RHS – rad hard software
Ring system – a disk of solid material around a planet.
RTC – real time clock.
RTG – Radioisotope Thermal Generator – electrical power plant
RTOS – real time operating system.
SDR – software defined radio
SDRAM – synchronous dynamic random access memory.
Segmentation – dividing a network or memory into sections.
Semiconductor – material with electrical characteristics between conductors and insulators; basis of current technology for processor, memory, and I/O devices, as well as sensors.
Semaphore – a binary signaling element among processes.
SD – secure digital (non-volatile memory).

SDVF – Software Development and Validation Facility.
Sensor – a device that converts a physical observable quantity or event to a signal.
Serial – bit by bit.
SEU – single event upset (radiation induced error).
Servo – a control device with feedback.
Six-pack – a six U Cubesat, 10 x 20 x 30 cm.
SMP – symmetric multiprocessing.
Snoop – monitor packets in a network, or data in a cache.
SN – (NASA's) Space Network
SOA – safe operating area; also, state of the art.
SOC – system on a chip; also state-of-charge.
Socket – an end-point in communication across a network
Soft core – a hardware description language description of a cpu core.
Software – set of instructions and data to tell a computer what to do.
SOI – Saturn Orbit insertion
Solar flare – a sudden rapid emission of electrons, ions, and atoms from a star.
Solar System – A star and its associated planets and such.
Solar wind – stream of charged particles emitted from a star's upper atmosphere.
SMP – symmetric multiprocessing.
Snoop – monitor packets in a network, or data in a cache.
Spacewire – high speed (160 Mbps) link.
SPI - Serial Peripheral Interface - a synchronous serial communication interface.
SRAM – static random access memory.
STAR – self test and repair.
State machine – model of sequential processes.
STOL – system test oriented language, a scripting language for testing systems.
Strawman – an early concept or prototype, to be refined.
SWAP – size, weight, power
T&I – test and integration.

Terrabyte – 10^{12} bytes.
SAA – South Atlantic anomaly. High radiation zone in Earth's atmosphere.
SEB – single event burnout.
SEU – single event upset.
SEL – single event latchup.
Soc – state of charge; system on a chip.
Soft core – hardware description description language model of a logic core.
SOI – silicon on insulator
SoS – silicon on sapphire – an inherently radiation-hard technology
spi – serial peripheral interface
SpaceCube – an advanced FPGA-based flight computer.
SpaceWire – networking and interconnect standard.
SRAM – static random access memory.
Stack – first in, last out data structure. Can be hardware or software.
Stack pointer – a reference pointer to the top of the stack.
State machine – model of sequential processes.
SWD – serial wire debug.
Synchronous – using the same clock to coordinate operations.
System – a collection of interacting elements and relationships with a specific behavior.
System of Systems – a complex collection of systems with pooled resources.
Suitsat – old Russian spacesuit, instrumented with an 8-bit micro, and launched from the ISS.
Swarm – a collection of satellites that can operate cooperatively.
sync – synchronize, synchronized.
TCP/IP – Transmission Control Protocol/Internet protocol.
TDRSS – Tracking and Data Relay Satellite System, Earth orbit.
Tera - 10^{12} or 2^{40}
Test-and-set – coordination mechanism for multiple processes that allows reading to a location and writing it in a non-interruptible manner.

Thread – smallest independent set of instructions managed by a multiprocessing operating system.
TID – total ionizing dose.
Tidal lock – where the same side of a object always faces the primary it is orbiting.
TMR – triple modular redundancy.
TNO – Trans-Neptunian objects.
Toolchain – set of software tools for development.
Transceiver – receiver and transmitter in one box.
Transducer – a device that converts one form of energy to another.
Train – a series of satellites in the same or similar orbits, providing sequential observations.
TRAP – exception or fault handling mechanism in a computer; an operating system component.
Triplicate – using three copies (of hardware, software, messaging, power supplies, etc.). for redundancy and error control.
TRL – technology readiness level
Trojan - minor planet that shares an orbit with one of the larger planets.
Truncate – discard. cutoff, make shorter.
TT&C – tracking, telemetry, and command.
UDP – User datagram protocol; part of the Internet Protocol.
Underflow – the result of an arithmetic operation is smaller than the smallest representable number.
Uplink – from ground to space.
USAF – United States Air Force.
USB – universal serial bus.
UV - ultraviolet
VDC – volts, direct current.
Vector – single dimensional array of values.
VHDL – very high level design language.
Virtualization – creating a virtual resource from available physical resources.
Virus – malignant computer program.
WiFi – short range digital radio.

Watchdog – hardware/software function to sanity check the hardware, software, and process; applies corrective action if a fault is detected; fail-safe mechanism.

Wiki – the Hawaiian word for "quick." Refers to a collaborative content website.

Word – a collection of bits of any size; does not have to be a power of two.

X-band – 7 – 11 GHz.

Zombie-sat – a dead satellite, in orbit.

Zone of Exclusion – volume in which the presence of an object personnel, or activities are prohibited.

Country's that have launched Cubesats

as of 9/18/21

This list was complied in September of 2021, and may not be complete. This lists the countrys that launched Cubesats as a part of their National Space Program, or by a University in the Country.

1. Australia
2. Bangladesh
3. Belgium
4. Brazil
5. Canada
6. China
7. Costa Rica
8. Denmark
9. Ecuador
10. Ethiopia
11. Finland
12. France
13. Galacia (autononous area of Spain)
14. Germany
15. Ghana
16. Greece
17. India
18. Italy
19. Japan

20. Jourdan
21. Kenya
22. Korea (south)
23. Lithuania
24. Malaysia
25. Mongolia
26. Netherlands
27. New Zealand
28. Nigeria
29. Norway
30. Pakistan
31. Peru
32. Poland
33. Puerto Rico
34. Romania
35. Singapore
36. Slovakia
37. South Africa
38. Spain
39. Switerland
40. Tunsia
41. Turkey
42. UK
43. Uruguay
44. USA
45. Vietnam

If you enjoyed this book, you might also enjoy one of my other books in the Space series.

Stakem, Patrick H. *16-bit Microprocessors, History and Architecture*, 2013 PRRB Publishing, ISBN-1520210922.

Stakem, Patrick H. *4- and 8-bit Microprocessors, Architecture and History*, 2013, PRRB Publishing, ISBN-152021572X,

Stakem, Patrick H. *Apollo's Computers,* 2014, PRRB Publishing, ISBN-1520215800.

Stakem, Patrick H. *The Architecture and Applications of the ARM Microprocessors,* 2013, PRRB Publishing, ISBN-1520215843.

Stakem, Patrick H. *Earth Rovers: for Exploration and Environmental Monitoring,* 2014, PRRB Publishing, ISBN-152021586X.

Stakem, Patrick H. *Embedded Computer Systems, Volume 1, Introduction and Architecture*, 2013, PRRB Publishing, ISBN-1520215959.

Stakem, Patrick H. *The History of Spacecraft Computers from the V-2 to the Space Station*, 2013, PRRB Publishing, ISBN-1520216181.

Stakem, Patrick H. *Floating Point Computation*, 2013, PRRB Publishing, ISBN-152021619X.

Stakem, Patrick H. *Architecture of Massively Parallel Microprocessor Systems*, 2011, PRRB Publishing, ISBN-1520250061.

Stakem, Patrick H. *Multicore Computer Architecture,* 2014, PRRB Publishing, ISBN-1520241372.

Stakem, Patrick H. *Personal Robots*, 2014, PRRB Publishing, ISBN-1520216254.

Stakem, Patrick H. *RISC Microprocessors, History and Overview,* 2013, PRRB Publishing, ISBN-1520216289.

Stakem, Patrick H. *Robots and Telerobots in Space Applications*, 2011, PRRB Publishing, ISBN-1520210361.

Stakem, Patrick H. *The Saturn Rocket and the Pegasus Missions, 1965,* 2013, PRRB Publishing, ISBN-1520209916.

Stakem, Patrick H. *Visiting the NASA Centers, and Locations of Historic Rockets & Spacecraft,* 2017, PRRB Publishing, ISBN-1549651205.

Stakem, Patrick H. *Microprocessors in Space*, 2011, PRRB Publishing, ISBN-1520216343.

Stakem, Patrick H. Computer *Virtualization and the Cloud*, 2013, PRRB Publishing, ISBN-152021636X.

Stakem, Patrick H. *What's the Worst That Could Happen? Bad Assumptions, Ignorance, Failures and Screw-ups in Engineering Projects, 2014,* PRRB Publishing, ISBN-1520207166.

Stakem, Patrick H. *Computer Architecture & Programming of the Intel x86 Family, 2013,* PRRB Publishing, ISBN-1520263724.

Stakem, Patrick H. *The Hardware and Software Architecture of the Transputer*, 2011,PRRB Publishing, ISBN-152020681X.

Stakem, Patrick H. *Mainframes, Computing on Big Iron*, 2015, PRRB Publishing, ISBN- 1520216459.

Stakem, Patrick H. *Spacecraft Control Centers*, 2015, PRRB Publishing, ISBN-1520200617.

Stakem, Patrick H. *Embedded in Space,* 2015, PRRB Publishing, ISBN-1520215916.

Stakem, Patrick H. *A Practitioner's Guide to RISC Microprocessor Architecture*, Wiley-Interscience, 1996, ISBN-0471130184.

Stakem, Patrick H. *Cubesat Engineering*, PRRB Publishing, 2017, ISBN-1520754019.

Stakem, Patrick H. *Cubesat Operations*, PRRB Publishing, 2017, ISBN-152076717X.

Stakem, Patrick H. *Interplanetary Cubesats*, PRRB Publishing, 2017, ISBN-1520766173 .

Stakem, Patrick H. Cubesat Constellations, Clusters, and Swarms, Stakem, PRRB Publishing, 2017, ISBN-1520767544.

Stakem, Patrick H. *Graphics Processing Units, an overview*, 2017, PRRB Publishing, ISBN-1520879695.

Stakem, Patrick H. *Intel Embedded and the Arduino-101, 2017,* PRRB Publishing, ISBN-1520879296.

Stakem, Patrick H. *Orbital Debris, the problem and the mitigation*, 2018, PRRB Publishing, ISBN-*1980466483*.

Stakem, Patrick H. *Manufacturing in Space*, 2018, PRRB Publishing, ISBN-1977076041.

Stakem, Patrick H. *NASA's Ships and Planes*, 2018, PRRB Publishing, ISBN-1977076823.

Stakem, Patrick H. *Space Tourism*, 2018, PRRB Publishing, ISBN-1977073506.

Stakem, Patrick H. *STEM – Data Storage and Communications*, 2018, PRRB Publishing, ISBN-1977073115.

Stakem, Patrick H. *In-Space Robotic Repair and Servicing*, 2018, PRRB Publishing, ISBN-1980478236.

Stakem, Patrick H. *Introducing Weather in the pre-K to 12 Curricula, A Resource Guide for Educators*, 2017, PRRB Publishing, ISBN-1980638241.

Stakem, Patrick H. *Introducing Astronomy in the pre-K to 12 Curricula, A Resource Guide for Educators*, 2017, PRRB Publishing, ISBN-198104065X.
Also available in a Brazilian Portuguese edition, ISBN-1983106127.

Stakem, Patrick H. *Deep Space Gateways, the Moon and Beyond*, 2017, PRRB Publishing, ISBN-1973465701.

Stakem, Patrick H. *Exploration of the Gas Giants, Space Missions to Jupiter, Saturn, Uranus, and Neptune*, PRRB Publishing, 2018, ISBN-9781717814500.

Stakem, Patrick H. *Crewed Spacecraft*, 2017, PRRB Publishing, ISBN-1549992406.

Stakem, Patrick H. *Rocketplanes to Space*, 2017, PRRB Publishing, ISBN-1549992589.

Stakem, Patrick H. *Crewed Space Stations,* 2017, PRRB Publishing, ISBN-1549992228.

Stakem, Patrick H. *Enviro-bots for STEM: Using Robotics in the pre-K to 12 Curricula, A Resource Guide for Educators,* 2017, PRRB Publishing, ISBN-1549656619.

Stakem, Patrick H. *STEM-Sat, Using Cubesats in the pre-K to 12 Curricula, A Resource Guide for Educators,* 2017, ISBN-1549656376.

Stakem, Patrick H. *Lunar Orbital Platform-Gateway,* 2018, PRRB Publishing, ISBN-1980498628.

Stakem, Patrick H. *Embedded GPU's,* 2018, PRRB Publishing, ISBN- 1980476497.

Stakem, Patrick H. *Mobile Cloud Robotics,* 2018, PRRB Publishing, ISBN- 1980488088.

Stakem, Patrick H. *Extreme Environment Embedded Systems,* 2017, PRRB Publishing, ISBN-1520215967.

Stakem, Patrick H. *What's the Worst, Volume-2,* 2018, ISBN-1981005579.

Stakem, Patrick H., *Spaceports,* 2018, ISBN-1981022287.

Stakem, Patrick H., *Space Launch Vehicles,* 2018, ISBN-1983071773.

Stakem, Patrick H. *Mars,* 2018, ISBN-1983116902.

Stakem, Patrick H. *X-86, 40th Anniversary ed,* 2018, ISBN-1983189405.

Stakem, Patrick H. *Lunar Orbital Platform-Gateway*, 2018, PRRB Publishing, ISBN-1980498628.

Stakem, Patrick H. *Space Weather*, 2018, ISBN-1723904023.

Stakem, Patrick H. *STEM-Engineering Process*, 2017, ISBN-1983196517.

Stakem, Patrick H. *Space Telescopes,* 2018, PRRB Publishing, ISBN-1728728568.

Stakem, Patrick H. *Exoplanets*, 2018, PRRB Publishing, ISBN-9781731385055.

Stakem, Patrick H. *Planetary Defense*, 2018, PRRB Publishing, ISBN-9781731001207.

Patrick H. Stakem *Exploration of the Asteroid Belt*, 2018, PRRB Publishing, ISBN-1731049846.

Patrick H. Stakem *Terraforming*, 2018, PRRB Publishing, ISBN-1790308100.

Patrick H. Stakem, *Martian Railroad,* 2019, PRRB Publishing, ISBN-1794488243.

Patrick H. Stakem, *Exoplanets,* 2019, PRRB Publishing, ISBN-1731385056.

Patrick H. Stakem, *Exploiting the Moon,* 2019, PRRB Publishing, ISBN-1091057850.

Patrick H. Stakem, *RISC-V, an Open Source Solution for Space Flight Computers,* 2019, PRRB Publishing, ISBN-1796434388.

Patrick H. Stakem, *Arm in Space*, 2019, PRRB Publishing, ISBN-9781099789137.

Patrick H. Stakem, *Extraterrestrial Life*, 2019, PRRB Publishing, ISBN-978-1072072188.

Patrick H. Stakem, *Space Command*, 2019, PRRB Publishing, ISBN-978-1693005398.

CubeRovers, A Synergy of Technologys, 2020, PRRB Publishing, ISBN-979-8651773138.

Robotic Exploration of the Icy moons of the Gas Giants. 2020, PRRB Publishing, ISBN- 979-8621431006

Hacking Cubesats, 2020, PRRB Publishing, ISBN-979-8623458964.

History & Future of Cubesats, PRRB Publishing, ISBN-979-8649179386.

Hacking Cubesats, Cybersecurity in Space, 2020, PRRB Publishing, ISBN-979-8623458964.

Powerships, Powerbarges, Floating Wind Farms: electricity when and where you need it, 2021, PRRB Publishing, ISBN-979-8716199477.

Hospital Ships, Trains, and Aircraft, 2020, PRRB Publishing, ISBN-979-8642944349.

2020/2021 Releases

CubeRovers, a Synergy of Technologys, 2020, ISBN-979-8651773138

Exploration of Lunar & Martian Lava Tubes by Cube-X, ISBN-979-8621435325.

Robotic Exploration of the Icy moons of the Gas Giants, ISBN-979-8621431006.

History & Future of Cubesats, ISBN-978-1986536356.

Robotic Exploration of the Icy Moons of the Ice Giants, by Swarms of Cubesats, ISBN-979-8621431006.

Swarm Robotics, ISBN-979-8534505948.

Introduction to Electric Power Systems, ISBN-979-8519208727.

Centros de Control: Operaciones en Satélites del Estándar CubeSat (Spanish Edition), 2021, ISBN-979-8510113068.

www.ingramcontent.com/pod-product-compliance
Lightning Source LLC
Chambersburg PA
CBHW020929180526
45163CB00007B/2943

www.ingramcontent.com/pod-product-compliance
Lightning Source LLC
Chambersburg PA
CBHW020929180526
45163CB00007B/2943